GEOGRAPHICAL
SKILLS AND FIELDWORK
AQA GCSE (9–1) GEOGRAPHY

Steph Warren

D1354860

Every effort has been made to trace all copyright holders, but if any have been inadvertently overlooked, the Publishers will be pleased to make the necessary arrangements at the first opportunity.

Although every effort has been made to ensure that website addresses are correct at time of going to press, Hodder Education cannot be held responsible for the content of any website mentioned in this book. It is sometimes possible to find a relocated web page by typing in the address of the home page for a website in the URL window of your browser.

Hachette UK's policy is to use papers that are natural, renewable and recyclable products and made from wood grown in sustainable forests. The logging and manufacturing processes are expected to conform to the environmental regulations of the country of origin.

Orders: please contact Bookpoint Ltd, 130 Park Drive, Milton Park, Abingdon, Oxon OX14 4SE. Telephone: (44) 01235 827720. Fax: (44) 01235 400454. Email education@bookpoint. co.uk Lines are open from 9 a.m. to 5 p.m., Monday to Saturday, with a 24-hour message answering service. You can also order through our website: www.hoddereducation.co.uk

ISBN: 978 1 4718 6590 9

© Steph Warren 2016

First published in 2016 by

Hodder Education,

An Hachette UK Company

Carmelite House

50 Victoria Embankment

London EC4Y 0DZ

www.hoddereducation.co.uk

Impression number 10 9 8 7 6 5

2022 2021 2020 2019

Cover photo © Frederic Bos-Fotolia.com

Illustrations by Gray Publishing and David Gardner

Typeset in AvenirLTStd-Light, 10/12 pts. by Aptara Inc.

Printed in Dubai

A catalogue record for this title is available from the British Library.

Contents

Acknowledgements

The publishers would like to thank the following for permission to reproduce copyright material:

Photo credits

p.1 © Steph Warren; **p.13** © Steph Warren; **p.23** *t* © Steph Warren; *b* Skyscan.co.uk/© GeoPerspectives; **p.24** *t* Courtesy of NASA-Johnson Space Center; *b* Skyscan.co.uk/© GetMapping plc; **p.25** © Steph Warren; **p.26** © Steph Warren; **p.27** © Historic England Archive (Aerofilms Collection); **p.29** *t* © Commission Air/Alamy Stock Photo; *b* © Steph Warren; **p.30** © Steph Warren; **p.67** all © Steph Warren; **p.70** Courtesy of Peak Electronic Design Limited; **p.72** *t* © Steph Warren *b* © GeographySouthWest/Moment/Getty Images; **p.74** © Steph Warren; **p.79** © GSO Images/The Image Bank/Getty Images; **p.82** *br* & **p.83** © Steph Warren; **p.85** © Steph Warren; **p.88** © Steph Warren; **p.89** © Steph Warren; **p.93** *t* Skyscan.co.uk/© GeoPerspectives; *b* Skyscan.co.uk/© GeoPerspectives; **p.96** © Steph Warren.

Maps on **pp.58, 60, 82, 84, 86–7, 90–1, 94–5, 97** are reproduced from Ordnance Survey mapping with permission of the Controller of HMSO. © Crown copyright. All rights reserved. Licence number 100036470.

Cartographic skills

n Chapter 1

Cartographic skills you will:

→ study human and physical features on atlas maps

→ study scale, symbols, grid references and direction on OS maps

→ study how to measure distances on OS maps

→ study landscape features and their characteristics on OS maps

→ study the physical and human landscape on OS maps

→ study OS map evidence of human activity including tourism

→ draw, label, understand and interpret sketch maps

→ study ground, aerial and satellite photographs

→ study the use of OS maps in association with photographs and sketches

→ draw, label and annotate sketches from photographs and in the field

→ draw, label and annotate diagrams and graphs.

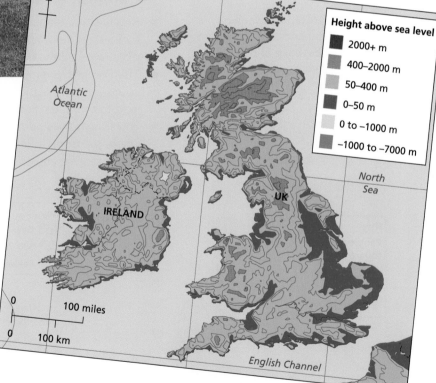

Height above sea level
- 2000+ m
- 400–2000 m
- 50–400 m
- 0–50 m
- 0 to –1000 m
- –1000 to –7000 m

N

Atlantic Ocean

IRELAND

UK

North Sea

English Channel

0   100 miles

0   100 km

1 Cartographic skills

Atlas maps

Atlases not only contain maps showing where places are, but also show physical patterns such as the height of the land and human patterns such as population density.

Physical features on atlas maps	Human features on atlas maps
Relief	Population distribution
Drainage	Population movements
	Transport networks
	Settlement layout

You will need to be able to describe these patterns and to understand how they relate to each other. Atlases contain maps at a variety of different scales, but the most common ones show patterns on a country, continent and world scale. You should be familiar with all of these.

How to describe the pattern of physical or human features on an atlas map

- Begin with a general statement about where the features are located on the map. For example, in Figure 1a China is densely populated on the eastern side of the country.
- Then go into greater detail such as mentioning the area of the country, or any particular features such as the name of the sea next to that area, in this case the Yellow Sea.
- You should then be more specific and include in the answer where the population is concentrated, such as the River Yangtze. In this section you could also give a reference using latitude and longitude lines.

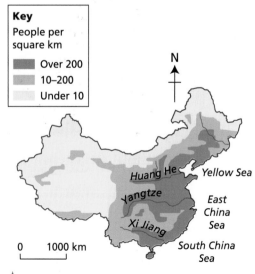

Figure 1a Population density of China

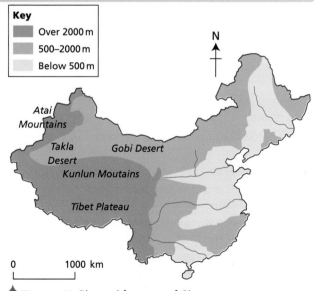

Figure 1b Physical features of China

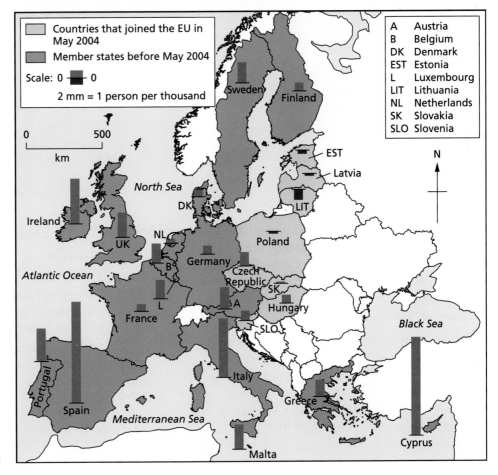

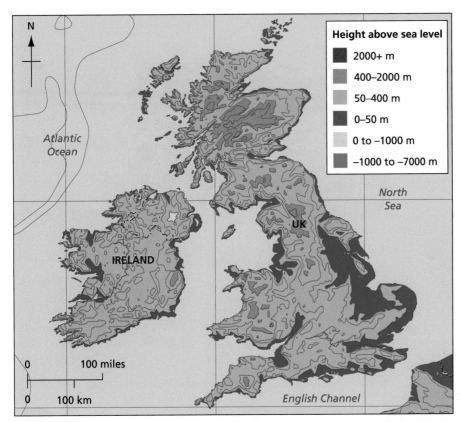

Height above sea level

- 2000+ m
- 400–2000 m
- 50–400 m
- 0–50 m
- 0 to −1000 m
- −1000 to −7000 m

N

Atlantic
Ocean

North
Sea

UK

IRELAND

0   100 miles

0   100 km

English Channel

◀ **Figure 2** A relief map of the
British Isles

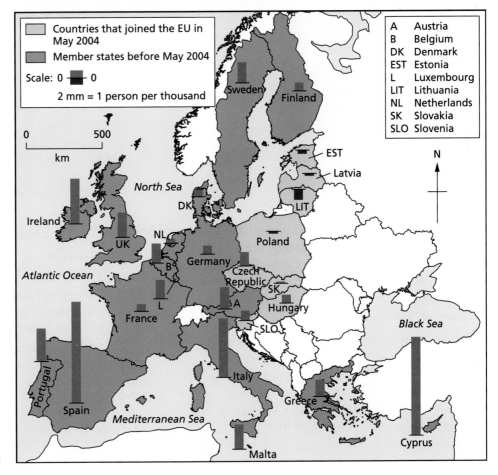

Countries that joined the EU in
May 2004

Member states before May 2004

Scale: 0 ━ ■ ━ 0

2 mm = 1 person per thousand

A	Austria
B	Belgium
DK	Denmark
EST	Estonia
L	Luxembourg
LIT	Lithuania
NL	Netherlands
SK	Slovakia
SLO	Slovenia

N

0   500

km

North Sea

Sweden

Finland

EST

Latvia

LIT

DK

Ireland

UK

NL

Poland

Atlantic Ocean

B

Germany

Czech
Republic

SK

L

A

Hungary

France

SLO

Black Sea

Portugal

Spain

Italy

Greece

Mediterranean Sea

Cyprus

Malta

➡ **Figure 3** Migration
change 2004–2005 in the EU

Latitude and longitude

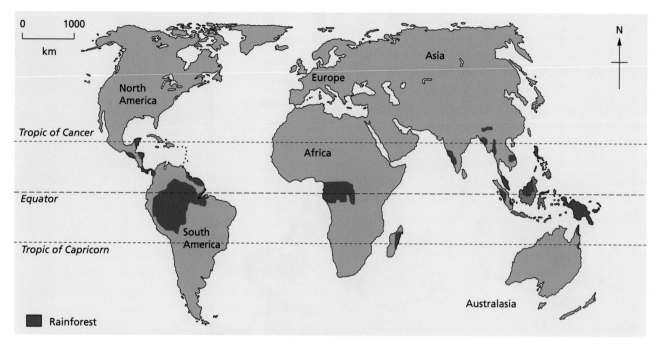

Figure 4 Tropical rainforests of the world

ACTIVITIES

a Describe the distribution of tropical rainforests shown in Figure 4.

Answer the following questions:

➤ Where are most of the rainforests in relation to the named latitude lines?
➤ Which continent has the largest area of rainforest?
➤ Which continent does not have any rainforests?
➤ Name a country where rainforests can be found.

Use questions like these to help you to answer the following questions:

b Describe the pattern of relief shown in Figure 2.
c What would you add to Figure 2 to make the description easier to complete?
d Describe the pattern of migration shown in Figure 3.

Latitude lines are the lines that go around the globe horizontally. The major lines of latitude are the Equator, Tropic of Cancer, Arctic Circle, Tropic of Capricorn and Antarctic Circle. These lines start at the Equator which is 0° and move north and south up to 90°N and 90°S respectively.

Longitude lines are lines that go vertically north and south on the globe. The major lines of longitude are the Greenwich Meridian and the International Date Line. These lines start at the Greenwich Meridian, which is 0°, and move east and west to the International Date Line, which is at 180°.

How to find a location using latitude and longitude

Latitude and longitude references are given in degrees and minutes from either the Equator or the Greenwich Meridian. The latitude reference is always given first followed by either N (north) or S (south). The longitude reference is then given followed by E (east) or W (west). A section of an atlas map for SW England is shown in Figure 5. The reference for Tiverton would be 50°54′N 3°29′W.

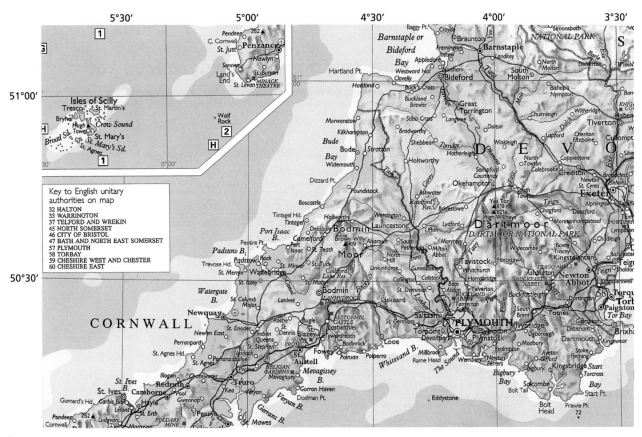

⬆ **Figure 5** An atlas map of part of the south-west of England

ACTIVITIES

Using Figure 5:
a Give the position of Exeter using latitude and longitude coordinates.
b Which settlement can be found at coordinates 50°22′N 4°10′W?
c Work in a pair. Make up four more questions for latitude and longitude. Your partner should answer them. Mark each other's work.

STRETCH AND CHALLENGE

• •

Explain the relationship between the human and physical features shown on Figure 1.

REVIEW

By the end of this section you should be able to:

✔ recognise and describe distributions and patterns of human features on atlas maps at a variety of scales
✔ recognise and describe distributions and patterns of physical features on atlas maps at a variety of scales
✔ analyse the relationship between physical and human factors on atlas maps.

Exam Tip

When describing physical and human features, always start with the general and finish with the specific.

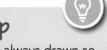
Ordnance Survey (OS) maps

Scale, symbols, grid references and direction

Scale

Maps are produced at different scales. The scale of the map is how much smaller the map is than the area it represents. On 1:50 000 scale maps, 1 cm on the map is 50 000 cm (500 metres) on the ground. On 1:25 000 scale maps, 1 cm on the map is 25 000 cm (250 metres) on the ground. The light blue lines which run horizontally and vertically across the map are called grid lines; they are drawn 2 cm apart on 1:50 000 maps and 4 cm apart on 1:25 000 maps. Every map has a scale line which can be used to measure distances.

Symbols

The symbols used on the map represent features on the ground. Many of the symbols give a 'clue' to what they represent. For example, a Post Office is the letter P. You will always be provided with a key but it is a good idea to familiarise yourself with OS maps so that you can interpret the symbols quickly without continual reference to the key.

An OS map symbols quick reference guide is provided below.

Feature	What to look for?
Woodland	Green areas with different types of trees identified
Water	Blue areas
Tourist information	These are symbols in blue
Fields	Usually white areas
Urban areas	Brown/beige shading
Roads	Motorways – blue A roads – green (major routes) and red (main roads) Secondary roads (B roads) – orangey/yellow Small side roads – yellow
Contours (showing height above sea level)	Thin brown lines. They are at 10 m intervals with every 50 m in a darker colour
Spot heights	These are a number next to a dot. They show the height of the land at the point where the dot is
Triangulation pillars	These are a number next to a symbol and also show the height at that point

Grid references

You will need to be able to locate features using four- and six-figure grid references. Four-figure grid references locate a square on the map and are usually used for large features such as an area of settlement or woodland. Six-figure grid references locate a particular point on the map such as a Post Office.

How to find a grid reference

Study the grid in Figure 6a. It has grid lines 2 cm apart and a number of symbols on it. To find the position of the Post Office in grid square 5885:

- Go along the bottom of the grid until you find the number 58.
- These are the first two numbers of the four-figure reference. The line will always be to the left of the symbol. Or if you were on the ground following the map, it would be the last grid line you walked over.
- Go up the side of the grid until you reach 85.
- These are the third and fourth numbers of the grid reference. The line will always be below the symbol you are locating. Or if you were on the ground following the map, it would be the last grid line you walked over.
- The symbol will be in the square right and above where these two lines meet.
- If you have been given a six-figure grid reference, the third and sixth numbers refer to the exact position of the symbol, for example 588857.
- You should complete the four-figure part of the reference to find the line and then divide the square into tenths in your mind, in this case, 8/10s as the reference is 588.
- Then complete the second part of the reference in the same way – 857.
- It is important to remember that for features such as a school, you will be given the word 'Sch' near a building. It is the centre of the building that is used for the six-figure grid reference.

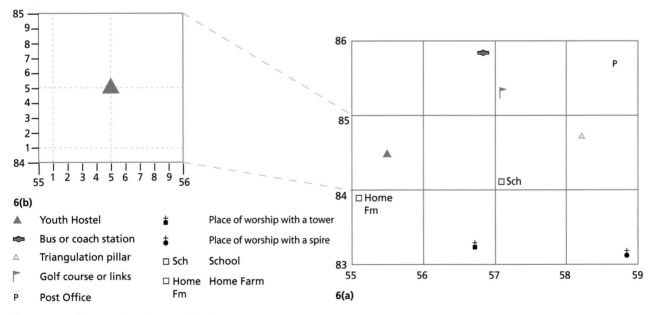

6(b)

▲   Youth Hostel

⬭   Bus or coach station

△   Triangulation pillar

⌐   Golf course or links

P   Post Office

⛫   Place of worship with a tower

⛫   Place of worship with a spire

☐ Sch   School

☐ Home Fm   Home Farm

6(a)

⮕ **Figure 6** Four-and six-figure grid references

How to give a grid reference

Study the grid in Figure 6a. It has lines 2 cm apart and a number of symbols on it. To give the position of the youth hostel:

- Give the number of the line immediately to the left of the symbol – it is 55.
- Give the number of the line immediately below the youth hostel symbol – it is 84.
- If a six-figure grid reference is required, you should divide the grid square in your mind into tenths as shown in Figure 6b. (This is double the size of the original square on the grid.) Each of the 1/10s is 2 mm because one grid square is 2 cm on the original grid.
- Therefore the four-figure grid reference for the youth hostel is 5584.
- The youth hostel is halfway across, 5/10s of the way or 555.
- The line below is 84. Again split the grid square into tenths, going upwards this time.
- The youth hostel is halfway up, 5/10s of the way or 845.
- The six-figure grid reference is written 555845.

Compass directions

Map directions are given using the standard 16-point compass as shown in Figure 7. In an exam, when an OS map is used, north will be taken as being at the top of the map following the grid lines.

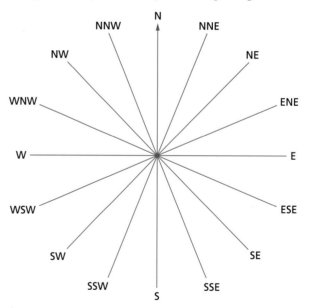

⬆ **Figure 7** A 16-point compass

ACTIVITIES

1 Copy and complete the following table using the key for OS maps on the front and back covers of the book.

Symbol	Feature
TH	
◥◤	
	Tunnel
⌃ ⌃ ⌃	
	Non-coniferous Woodland
CH	

2 Copy and complete the table below using Figure 6a and 6b.

Symbol	Key	Four-figure reference	Six-figure reference
✝■			
P			
△			
	Golf course or links		
□ Home Fm			
	Bus or coach station		
□ Sch			

Warkworth map extract (page 84)

3 Copy and complete the table below by drawing the symbol for flat rock found in grid square 2511.

Feature	Symbol
Flat rock	

4 Who owns the land in grid square 2508?

5 What type of woodland is found in 2007?

6 Identify three services in the village of Shilbottle in grid square 1908.

7 Complete the table by identifying five tourist features and their grid references found in Alnwick (north-west of the map extract). The first one has been done for you.

Tourist feature	Grid reference
Information	187133

8 Which direction is Alnwick from Alnmouth?

9 Which direction is Warkworth from Alnmouth?

10 Which direction is Shilbottle from Lesbury?

STRETCH AND CHALLENGE

1 Study Figure 24. The photograph was taken at grid reference 267048. In which direction was the camera pointing?

2 Look at Figure 23. Which direction was the camera pointing when the aerial photograph was taken?

REVIEW

By the end of this section you should be able to:

✔ understand what is meant by the scale of a map

✔ recognise symbols

✔ use four- and six-figure grid references accurately

✔ demonstrate an understanding of direction using the 16-point compass.

Exam Tip

Grid references and directions are often used relating to the location of photographs or features on a map.

Straight line and winding distances

Measuring straight line distances

When measuring straight line distances, it is useful to remember that on maps the grid lines are always drawn so that they represent 1 km on the ground. This can make measuring straight line distances very easy.

How to measure a straight line distance between two points such as two railway stations

- Use a ruler or the edge of a piece of paper.
- Mark on the piece of paper the location of the two points or take the measurement between the points with the ruler.
- Measure this distance against the scale line on the map or calculate the distance using the scale of 2 cm = 1 km.

For example, on the OS map of Swanage on page 97, the straight distance from the station at Corfe Castle to the station at Swanage is 14.6 cm or 7.3 km.

Measuring curved (winding) distances

Measuring the distance along a curved or winding route such as a road is more complicated. This can be done either by using a piece of string or by splitting the route into straight sections. The easiest way of measuring the distances along a winding route is with a piece of string.

How to measure a route (winding distance such as a railway line, road or river) using a piece of string

- The end of the piece of string should be placed at the beginning of the route.
- The string should be laid along the route following the curves as accurately as possible.
- The end of the route should be marked on the piece of string.
- The length of the string which equals the route can then be measured against the scale on the map or a calculation can be done on the basis that 2 cm = 1 km.

How to measure a route (winding distance such as a railway line, road or river) using straight sections

- The route to be measured should be split into a number of straight sections.
- The point chosen for the end of one straight section and the beginning of another is usually where the route bends.
- All of the straight sections can then be measured and the total distance converted into kilometres using the scale.
- The more straight sections that the route is broken down into, the more accurate will be the final measurement.

For example, Figure 8 is a section of the railway line on the OS map of Swanage on page 97. Notice how the line has been broken into straight sections.

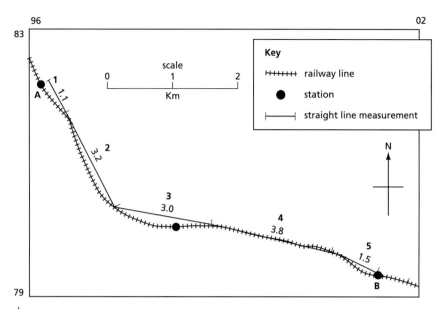

Figure 8 Measuring distance on an OS map

ACTIVITIES

Wensleydale map extract (page 86)

1 Calculate the length of the River Ure from the bridge at Hawes (8790) to the bridge at Bainbridge (9390).
2 What is the approximate length of the A road between 875898 and 934904?
3 Which is the longest route?
4 What is the difference in kilometres?
5 How far is it from one of the sources of Bardale Beck at 865838 to its confluence with Raydale Beck at 911861?

Swanage map extract (page 95)

6 What is the length of the B3069 between its two junctions (9681 and 0078) with the A351?
7 Which is the furthest distance, travelling between these two points on the A351 road or the B3069 road?
8 How much further is it?

New Forest (Bournemouth and Poole) map extract (page 87)

9 What is the length of the railway line on the map extract?
10 What is the length of the River Stour between grid reference 133940 and 150922?

Exam Tip

To make measuring distances easier, always have a piece of string in your pencil case.

REVIEW

By the end of this section you should be able to:

✔ measure straight line distances on a map
✔ measure curved distances on a map.

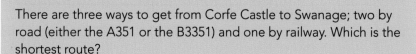

STRETCH AND CHALLENGE

There are three ways to get from Corfe Castle to Swanage; two by road (either the A351 or the B3351) and one by railway. Which is the shortest route?

Suggest two reasons for taking the longest route to do the journey.

How to construct a cross-section along the line X-Y in Figure 9

- Place the edge of a piece of paper along the line X-Y in diagram A.
- Mark the heights of the contour lines where the paper crosses them. This is usually done by putting a small line on the piece of paper and noting down the height at that point. Also mark the spot height which shows the top of the hill.
- Remove the piece of paper with the location of the contour lines and the heights marked on it. Place it on a piece of graph paper. Work out an appropriate scale for the y-axis and draw it on the graph paper. Mark points on the graph at the correct height and at the correct location as shown on diagram B. Point A, for example, is marked at 200 metres.
- You may draw lines up to the points as shown on diagrams B and C but it is perfectly correct to just draw the crosses at the correct point.
- Join up the crosses to show the shape of the land. In this case it is a hill with a steep slope and a less steep slope.

Landscape features and their characteristics

Cross-sections

A cross-section shows the variations in relief along a chosen line. It is a graph which shows distance along the x-axis (horizontal) and height on the y-axis (vertical). When drawing a cross-section, the scale used on both axes must be chosen carefully to show a true representation of the area.

Diagram A

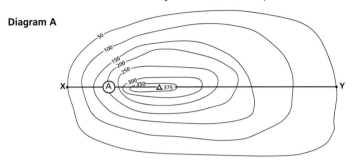

Diagram B

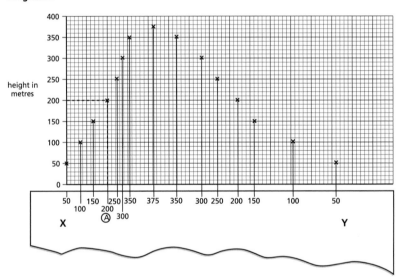

Diagram C

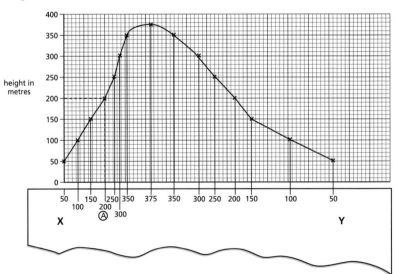

⬆ **Figure 9** Constructing a cross-section

Interpreting cross-sections

Cross-sections show the shape of the land. If the land is steep, the contours will be close together on the map and the line on the graph paper will increase quickly. If the land slopes gradually, the contours will be further apart and the line on the graph paper will increase more slowly. This can be seen in Figure 9. The first slope to the left of diagram A has a steep gradient with the contours on diagram A being quite close together. The contours to the right of diagram A are far apart depicting a gentle gradient that is reflected in the line of the graph on diagrams B and C.

Annotating cross-sections

Cross-sections show many physical features. As stated above, the most important feature is the shape of the land. Other physical features such as rivers can also be marked. Human features such as settlements and roads can also be located on a cross-section with explanatory points being made.

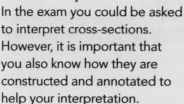

Exam Tip

In the exam you could be asked to interpret cross-sections. However, it is important that you also know how they are constructed and annotated to help your interpretation.

Exam Tip

Make sure that you know what the following terms mean:

Gradient – the slope of the land.

Contours – the lines on a map that show height.

Spot height – the height of the land at a particular point on the map.

You could be asked to define them or just asked questions about maps that include these terms.

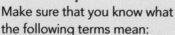

ACTIVITIES

Wensleydale map extract (page 86)

1 a Draw a cross-section from the 560 m contour line at Green Side (865851) to the 490 m contour line at New Close (884844).
   b Label the two rivers.
   c Mark on the cross-section where you think the photograph shown in Figure 10 was taken.
   d In which direction do you think the photograph was taken?
   e Justify your answer with reference to the photograph and the map.

⬆ **Figure 10**

2 Draw a cross-section from spot height 511 m in grid square 8987 to spot height 379 m in grid square 9487.
3 Label four secondary roads less than 4 m wide, three streams, Semer Water and two areas of woodland.

STRETCH AND CHALLENGE

Using the Wensleydale map extract, draw a cross-section from spot height 614 m in grid square 8786 to spot height 535 m in grid square 8892. Label the steepest slopes, the River Ure and the A684.

REVIEW

By the end of this section you should be able to:

✔ construct, interpret and annotate cross-sections and transects of physical and human landscapes
✔ use and understand gradient, contour and spot height.

⭐ Learning objective

– To study the physical and human landscape.

⭐ Learning outcomes

→ To be able to identify and describe basic landscape features.
→ To be able to identify major relief features on maps.
→ To be able to recognise and describe the patterns made by physical features such as relief (slopes), drainage (rivers and their valleys) and different types of vegetation.
→ To be able to describe the physical features as they are shown on large-scale maps of the following landscapes – coastlines, fluvial and glacial landscapes.
→ To be able to recognise and describe the distribution and patterns made by human features such as land use (settlement patterns) and communications.
→ To be able to compare maps.

Exam Tip

You will only need to be able to interpret two of the three landscapes: fluvial, coastal or glacial.

Interpreting the physical and human landscape

There are a number of physical and human features which you could be asked to recognise on OS maps, and you could be asked to describe the patterns they produce. It is important that you understand what the question expects you to write about.

The physical features that you could be asked about are:

Relief – the shape of the land. This can be seen by looking at the distance between the contours and the pattern that the contours show on the map. The height difference between each contour line (the vertical interval) is 10 m on 1:50000 maps. Every 50 m, the contour line is a slightly darker brown and has the height marked on it as shown in Figure 11 on page 15. On 1:25000 maps the vertical interval is 5 m with a darker contour line every 25 m. When writing about the relief you should use actual figures from the map, as well as terms such as highland, lowland or steep slopes. These are meaningless without the actual height or the rate at which the height is increasing or decreasing.

Fluvial landscapes (rivers and their valleys) – the skills you require are to be able to:

- recognise the presence or absence of rivers or lakes in an area
- recognise the patterns that rivers make in an area
- recognise the direction that a river is flowing
- see the influence of human activity on rivers and their valleys
- describe a river and its valley
- compare river valleys on a map extract
- compare river valleys on maps of differing scales.

Coastal landscapes – the skills you require are to be able to:

- recognise coastal features such as cliffs and stacks
- see the influence of human activity on coastal landscapes
- describe coastal landscapes
- compare coastlines on a map extract
- compare coastal landscapes on maps of differing scales.

Glacial landscapes – the skills you require are to be able to:

- recognise glacial features such as corries and glacial troughs
- see the influence of human activity on glacial landscapes
- describe glacial landscapes
- compare glacial valleys on a map extract
- compare glacial landscapes on maps of differing scales.

Vegetation – you will need to be able to recognise the type of vegetation, for example woodlands, orchards and parks, and describe their distribution on a map. These can be seen in the land features section of the key.

Contour patterns	Commentary
	This is a uniform slope. The contours decrease evenly. The distance between the contour lines is the same all the way down the slope.
	This is a convex slope. The contour lines are closer together at the bottom of the slope. This means that the height is increasing quickly because the slope is steep. The lines gradually become further apart towards the top of the slope. This means that the height is increasing more slowly and that the slope is more gentle.
	This is a concave slope. The contour lines are further apart at the bottom of the slope. This means that the height is increasing slowly and that the slope is gentle. The lines gradually become closer together towards the top of the slope. This means that the height is increasing more quickly and that the slope is steeper.
	This is a V-shaped valley. The distance between the contours is regular, meaning that the valley sides will have an even slope. There is no flat land by the side of the river. This is shown by the presence of contour lines right next to the river. The V of the contour always points upstream. In this instance, the river is flowing from the top to the bottom of the page.
	This is a U-shaped valley. The distance between the contours is regular, meaning that valley sides will have an even slope. The contours increase quickly, meaning that the sides of the valley are steep. There is an area of flat land by the side of the river shown by the lack of contour lines. The contour lines do not cross the river for many kilometres; it is therefore very difficult to ascertain the direction of flow. On an OS map, other clues would be the presence of tributaries and the river becoming wider as it moves downstream.

Exam Tip

Remember that contour lines never touch or cross.

- The V of the contours across a river always points upstream.
- Contour lines which are close together mean steep slopes.
- Contour lines which are far apart mean gentle slopes.

◄ **Figure 11** Basic contour patterns of slopes and valleys

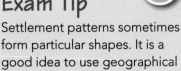

Exam Tip

Settlement patterns sometimes form particular shapes. It is a good idea to use geographical terminology to describe these shapes, such as:

Dispersed – this is when the settlements/houses are spread out over an area.

Nucleated – this is when settlements are close together in an area.

Linear – this is when houses in a settlement are either side of a main road.

The human features that you could be asked about are:

Land use – this would include urban areas and settlement patterns in rural areas. Typical questions would ask about the settlement distribution of an area and the human or physical factors which might have influenced the pattern.

Communications – this would include routes such as roads and railway lines. You might be asked to comment on the pattern shown by the communication links or explain how relief has influenced the pattern.

How to describe the pattern or distribution of settlement on a map

- Begin with a general statement about where settlement is located on the map, such as 'in the south-east'.
- Then go into greater detail, such as 'the settlement is along the lines of communication and at the bottom of a steep slope'. This mentions both a physical and human feature, although this is not essential.
- You should then be more specific and include in the answer the name of the road and the height of the land.

For example, there follows a description of the settlement pattern on the Wensleydale OS map (page 89). The majority of the settlement can be found in a band across the centre of the map extract. The settlements are close to the River Ure. The largest settlement is Hawes in grid square 8789. The settlements are mainly found on the south side of the valley, an exception being Askrigg in grid square 9491.

ACTIVITIES

Looe map extract (page 90)

1 Describe the distribution of woodland.
2 Describe the distribution of communication links in the area.

Swanage map extract (page 95)

3 Describe the distribution of settlement.
4 Woodland is being managed in the area of the map north of grid line 82. How does the map show that the woodland is being managed?
5 How have communication links been affected by the relief of the area?
6 Study the three map extracts for Swanage (pages 94, 95 and 97).
  a Which is the largest scale map? Comment on your answer. Ensure that you include how the maps differ in the detail that they show.

b Complete the table to show features that are only shown on that scale of the map.

1:50 000	1:25 000	1:10 000

c Which scale of map would be the best to use on a geography field trip?

d State two different ways in which each map can be used.

New Forest (Bournemouth and Christchurch) map extract (page 87)

7 What is the coastal landform found in grid square 1891?

8 Describe the coastal landscape found in grid square 1790.

9 Compare the coastal landscape found in grid square 1690 with that in 1491.

10 What is the fluvial feature found in grid square 1492?

11 Describe the fluvial landscape between grid reference 133940 and 140934.

STRETCH AND CHALLENGE

• •

Explain the communication links shown on the Cambridge OS map extract on page 91.

REVIEW

By the end of this section you should be able to:

✔ identify and describe basic landscape features

✔ identify major relief features on maps

✔ recognise and describe the patterns made by physical features such as relief (slopes), drainage (rivers and their valleys) and different types of vegetation

✔ describe the physical features as they are shown on large-scale maps of the following landscapes – coastlines, fluvial and glacial landscapes

✔ recognise and describe the distribution and patterns made by human features such as land use (settlement patterns) and communications

✔ compare maps.

Human activity from map evidence

Maps show a large amount of human activity, the extent of which is shown by looking at the key. Communication routes such as roads and railways with their associated features such as tunnels and level crossings (LC) are shown. Other public rights of way such as footpaths and cycle networks are also shown on maps.

The section on the OS map key called 'Land features' contains a lot of information about the location of other human activity, for example windmills and wind generators. It also indicates who owns some of the land on the map, such as the National Trust.

Maps show settlements but not the number of people who live in them, although their actual size is shown by the area they cover. Some human activity is shown by the services available in settlements, such as churches, Post Offices and public houses.

Tourist information has its own section in the key on OS maps and many symbols are used to depict the variety of tourist information that is available on maps. As well as the information in the key, it is also important to remember that castles (in the antiquities part of the key) are also tourist attractions.

ACTIVITIES

Looe map extract (page 90)

1   Duloe has five services. State what the services are by drawing the appropriate symbol.
2   Figure 12 on page 19 is a sketch map of part of the Looe map extract. Copy and complete the sketch map with the following features:
   a   the West Looe River
   b   the settlement of Looe
   c   the route of the A387
   d   100 m contour line
3   Copy and complete the table using the sketch map and the OS map extract of Looe to help you.

Place	Feature on map
Island A	
Spot height B	
Tourist attraction C	
Tourist attraction D	
Road number E	
Road number F	
Settlement G	
Type of woodland H	
Farm I	

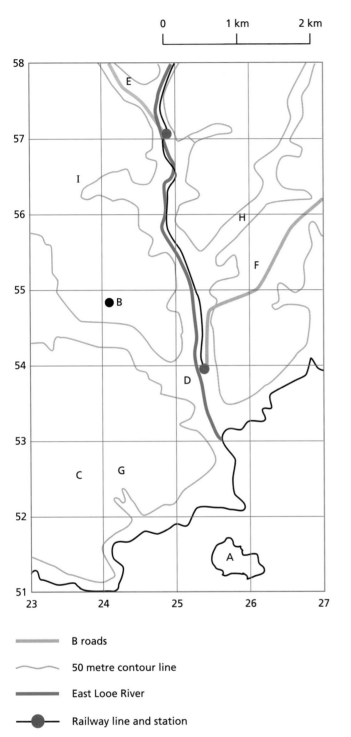

0   1 km   2 km

Legend:
- B roads
- 50 metre contour line
- East Looe River
- Railway line and station

Figure 12 An incomplete sketch map of part of the Looe extract

Exam Tip

Human activity may be examined by using a sketch map. On exam papers, it is better to shade rather than use different colours on sketch maps.

STRETCH AND CHALLENGE

Explain the presence of so many tourist information features on the Looe map extract on page 90.

REVIEW

By the end of this section you should be able to:

✔ interpret map evidence of human activity
✔ interpret map evidence of tourism.

Sketch maps

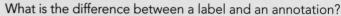

Exam Tip

What is the difference between a label and an annotation?

- A label is a simple descriptive point.
- An annotation is a label with more detailed description or an explanatory point.

How to draw, label, understand and interpret sketch maps

Many students are daunted by the idea of drawing sketch maps. What must be remembered is that a perfect replica of the map is not being expected, just a simple representation of the main features. There is an example of a sketch map on page 19, showing part of the Looe OS map from page 90.

During an exam, you may be asked to draw a sketch map of a particular feature, for example the woodland or the relief and drainage of an area. You must not redraw the whole of the map extract. What you must do is to show all of the features you have been asked for, plus one or two other distinctive features. If you are asked to draw the relief of an area you could simply shade in the area of land above a certain height. The sketch should be clear and well labelled or annotated depending on the command word in the question.

You should also be able to interpret a sketch map if you are asked questions about one in the examination. This will involve describing and explaining the physical and human geography which can be seen on the sketch map. It is important when interpreting photographs that your writing is coherent and shows good literacy skills in expressing geographical points. The sketch map will be like the original but will contain less detail, therefore you will need to be able to add to the detail in your written answer by interpreting what is there and filling in what has been omitted. For example, height of the land might only be shaded above a certain contour. But the land below that level could also be included in your answer.

How to draw a sketch from a map

- Using a ruler and pencil draw a frame to the size you want the sketch map to be. This can be the same size as the map or not, but a scale must be included.
- Lightly draw on the grid lines, writing their numbers down the side and across the bottom of the sketch map. These will act as guidelines.
- Draw in the features that are relevant to the question.
- Do not make the sketch map too detailed. It is not necessary to draw every feature, although you may wish to add other features besides the ones indicated in the question. For example, if you were asked to sketch the distribution of settlement in an area, it would be a good idea to include the road network.
- Add appropriate labels and annotations.
- You may wish to use colour. However, in the exam you will be required to only use a pencil/black pen. Therefore, it would be a good idea to practise completing sketch maps without using colours to differentiate between features.

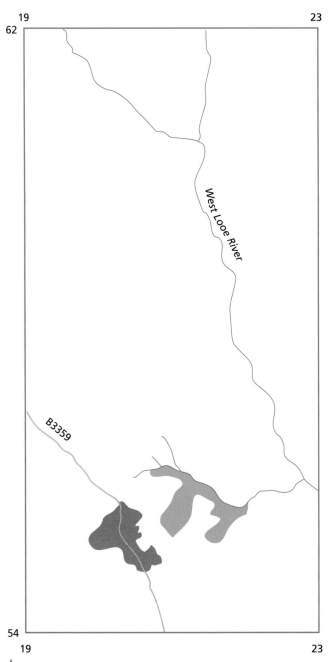

Figure 13 An incomplete sketch map of part of the Looe extract

ACTIVITIES

Looe map extract (page 90)

1 Trace Figure 13. Use it to complete a sketch map of the woodland west of easting 23.

Warkworth map extract (page 84)

2 Using the OS map, draw a sketch map of the area between eastings 23 to 27, northings 05 to 12. Mark the following on your sketch:

  a The built-up areas of Alnmouth and Warkworth

  b The Rivers Aln and Coquet

  c The coastline and all coastal features

  d Any tourist information features in the area

REVIEW

By the end of this section you should be able to:

✔ draw, label, understand and interpret sketch maps.

STRETCH AND CHALLENGE

Cambridge map extract (page 91)

Using the OS map, draw a sketch map for the area between eastings 46 to 51 and northings 64 to 67. Include all the human and physical features you feel are relevant to show what the area is like.

Ground, aerial and satellite photographs

How to interpret aerial, oblique and satellite photographs

Ground photographs are, as the name suggests, ones that are taken when the person is standing on the ground. They can be taken from many different angles and heights but the person is always on the ground. These photographs could be taken of landforms, natural vegetation, land use or settlement.

Aerial photographs can be taken vertical (directly above). If so, they are called 'bird's eye view' photographs because they show a picture taken as if looking down like a flying bird. These are particularly useful when trying to interpret land-use patterns of an area (see Figure 16). Other aerial photographs are oblique (taken from an angle) and therefore show more detail such as the sides of buildings. Figure 14 below shows the difference between vertical and oblique photographs.

A satellite image is a picture of Earth taken from space. Satellite images can show patterns on a large scale such as the lights from urban areas on a continent, or can zoom in to see small details such as cars on a street.

All of these different types of photographs show elements of the landscape which are not found on OS maps, such as types of crops being grown and the different uses of buildings.

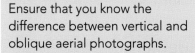

Exam Tip

Ensure that you know the difference between vertical and oblique aerial photographs.

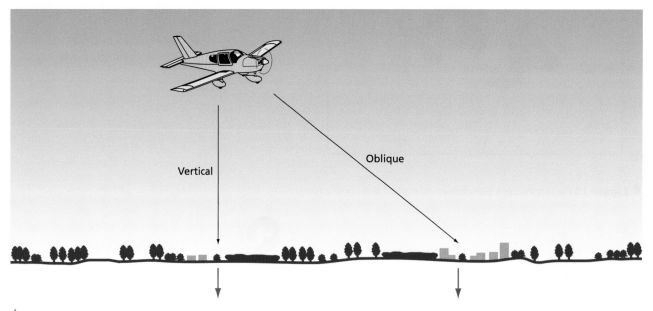

⬆ **Figure 14** The difference between vertical and oblique aerial photographs

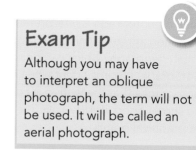
⬆ **Figure 15** Oblique aerial photograph of Swanage

⬆ **Figure 16** Vertical aerial photograph of Swanage

Interpretation of aerial photography

When a photograph is interpreted, it involves describing and explaining the physical and human geography which can be seen on the photograph. It is important when interpreting photographs that the writing is coherent and shows good literacy skills in expressing geographical points.

➡ **Figure 17** A satellite photograph of the Swanage area

Example of an interpretation of an aerial photograph

Physical features – relief

The land is very flat with no visible slopes. There are no rivers visible on the photograph. The soil is fertile, therefore a wide variety of crops can be grown.

Human features – land use

Farmland: The fields are large if they are compared with the individual farm buildings shown on the photograph. They are mostly square and rectangular. This implies the use of large arable machinery. It is arable land with a wide variety of crops being grown; this can be seen by the different colours. There are also lines in the fields which show that machinery has been used.

Because crops are growing in most of the fields, the photograph must have been taken in the spring or summer. Most of the fields are divided by roads and ditches with the only exception being tree boundaries in the north-east of the photograph which might be being used as a wind break to protect the village of Landbeach.

Settlement: The village of Landbeach, which is clearly shown to the east of the photograph, is a linear settlement. For the length of the village there is only one house on each side of the road (the road number can be added by reference to the OS map). To the west on the map, there are farms which display a dispersed land-use pattern.

➡ **Figure 18** An aerial photograph of Landbeach near Cambridge and its surroundings

ACTIVITIES

1 Study the different photographs of Swanage shown in Figures 15, 16 and 17 and Figure 20 on page 26. What do the different photographs show you? It may help to put your answer in a table like the one below:

Ground photograph	Oblique aerial photograph	Vertical aerial photograph	Satellite photograph

2 Study the aerial photograph of Warkworth in Figure 23 on page 29. Describe the human and physical features of the area.

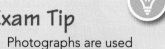
Figure 19 An oblique aerial photograph of Gatwick Airport

Exam Tip

- Photographs are used in exams to help you to identify features.
- You may also be asked to draw a sketch from a photograph.

STRETCH AND CHALLENGE

Study Figure 19 above, an aerial photograph of Gatwick Airport.

Draw an annotated sketch of the photograph.

REVIEW

By the end of this section you should be able to:

✔ interpret ground, and vertical and oblique aerial photographs
✔ interpret satellite photographs.

⭐ **Learning objective**

— To study the use of maps in association with photographs and sketches.

⭐ **Learning outcomes**

➜ To be able to interpret maps and relate them to photographs.
➜ To be able to interpret maps and relate them to sketches.

Use maps in association with photographs and sketches

Many exam papers have questions which require you to be able to use the OS map with a photograph. The types of questions you could be asked are:

■ recognition of certain features which have been identified by a letter on the photograph
■ the direction the photograph was taken
■ a comparison of features that can be seen on the map and not on the photograph and on the photograph and not on the map
■ the location of where the photograph was taken.

In order to be able to use a photograph with a map, you will have to orientate the photograph so that the features on the photograph are in the same position as those on the map. The map always has north at the top, a photograph may not! The way to do this is to look for an important feature on the photograph, such as the shape of the coastline on the Swanage map (page 95) and Figure 20. Then look at the town shown by the brown shading on the map and the houses on the photograph. It is obvious that the photograph is taken from the north of the map facing south. The exact location is worked out by looking at the fine detail of the photograph in relation to the map; in this case it would be the angle of the coastline, the position of the pier and the headland in the distance. Figure 20 has been annotated with the answers to questions that you might be asked when using a map with a photograph.

Other examination questions may require you to use the OS map to complete a sketch. There also could be sketches drawn with features labelled that you then have to identify using the map. There is an example of this on page 21 where there is a sketch of the Looe map. This has been covered in more detail earlier in this chapter.

Peveril Point
– a headland

Swanage town

New Swanage

Coastal defence on the beach: groynes

Swanage Bay

South West Coast Path

⬆ **Figure 20** A photograph of Swanage Bay taken from 040814 facing south

ACTIVITY

Looe map extract (page 90)

Study Figure 21. It is an aerial photograph of Looe.

a In which direction was the camera pointing?

b Copy and complete the table below using the photograph and the OS map of Looe on page 90 to help you.

Letter on photograph	Grid reference	Feature
V	Six figure =	Number of road =
W	Four figure =	Name of woodland =
X	Six figure =	Identify symbol =
Y	Four figure =	Name of river =
Z	Six figure =	Identify symbol =

c State two features that can be seen on the photograph but not on the map.

d State two features that can be seen on the map but not on the photograph.

Figure 21 Aerial photograph of Looe

REVIEW

By the end of this section you should be able to:

✔ interpret maps and relate them to photographs

✔ interpret maps and relate them to sketches.

Drawing sketches from photographs

Earlier in this chapter you learnt the difference between a label and an annotation and how to draw sketches from maps. In this section you will learn how to apply your knowledge to sketches. To draw a geographical sketch from a photograph you do not have to be an excellent artist, you just need to follow some basic steps.

How to draw a sketch from a photograph

■ Draw a frame to the size you want the sketch to be.
■ Lightly draw lines dividing the frame into four quarters. These will help you to draw the rest of the sketch, acting as guidelines. The lines can be erased when the sketch is completed.
■ Draw in the most important lines, such as rivers, coastline and the outline of the hills.
■ Draw in the less important features, such as woodland, settlements and communication lines. Do not make the sketch too detailed. It is not necessary to draw every feature.
■ Add appropriate labels and annotations.
■ Rub out the lightly drawn lines that divided the sketch to start off.

The sketch in Figure 22 is a drawing of Warkworth taken from the aerial photograph in Figure 23. It has been given descriptive labels (in blue) and annotations (in red) explaining its site.

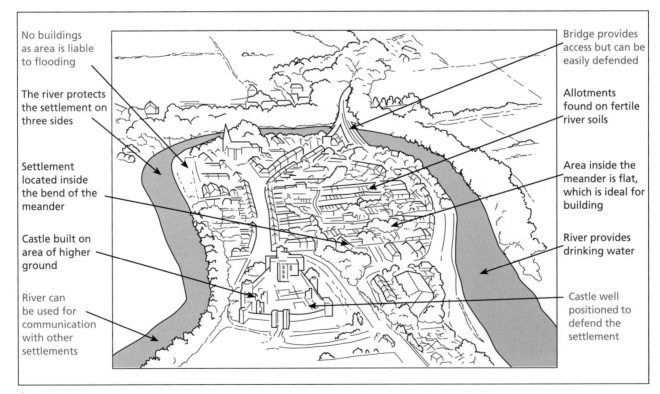

No buildings as area is liable to flooding

The river protects the settlement on three sides

Settlement located inside the bend of the meander

Castle built on area of higher ground

River can be used for communication with other settlements

Bridge provides access but can be easily defended

Allotments found on fertile river soils

Area inside the meander is flat, which is ideal for building

River provides drinking water

Castle well positioned to defend the settlement

⬆ **Figure 22** Sketch of Warkworth

⬆ **Figure 23** Aerial photograph of Warkworth

⬆ **Figure 24** Photograph of Warkworth from Amble

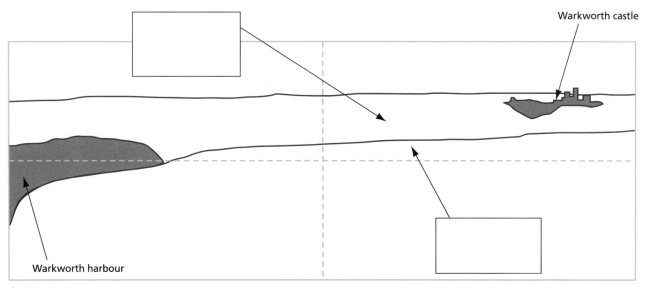

Warkworth castle

Warkworth harbour

⬆ **Figure 25** Incomplete sketch of Warkworth from Amble

⬆ **Figure 26** Photograph of St Bees in Cumbria

ACTIVITIES

1   Study Figure 24. It is a photograph of Warkworth from Amble. Trace Figure 25 – an incomplete sketch of the photograph. Complete the sketch as directed in the 'how to' box on page 28. The boxes give ideas of where you could add labels.
2   Draw a sketch of Figure 26. Label the following on the sketch:
    a   A caravan site
    b   Types of coastal defence
    c   The beach
    d   The cliffs
    e   The village of St Bees
    f   The Lake District hills in the background

STRETCH AND CHALLENGE

• •

On your sketch of Warkworth (from question 1 of the Activities), add some annotations to explain the site of Warkworth.

Exam Tip

In exams, you could be asked to complete sketches rather than draw them from scratch.

REVIEW

By the end of this section you should be able to:

✔ draw sketches from photographs and in the field
✔ label and annotate sketches.

Label and annotate diagrams and graphs

All geography students need the basic skills of being able to draw, label and annotate diagrams and graphs. These skills will be required to achieve good grades in all geography exams.

Diagrams

Diagrams should be used in geography to demonstrate knowledge. They can be particularly useful where you are required to explain the formation of landforms. For example, the diagram in Figure 27 shows how cliffs retreat. It can sometimes be more appropriate to draw a series of diagrams to show the sequence of something happening.

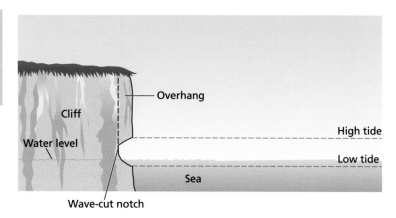

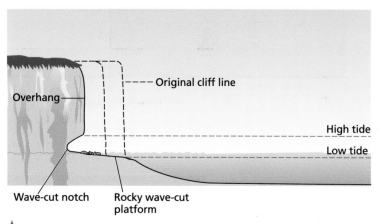

⬆ **Figure 27** Cliffs and wave-cut platforms

Graphs

Graphs can be labelled to point out changes and patterns, and annotated to explain what is happening. How to draw different graphs is dealt with in Chapter 2 Graphical skills. The population pyramid in Figure 28 has been labelled on the left-hand side to point out the important features. The right-hand side contains annotations which develop the points a little further and therefore are not simple labels.

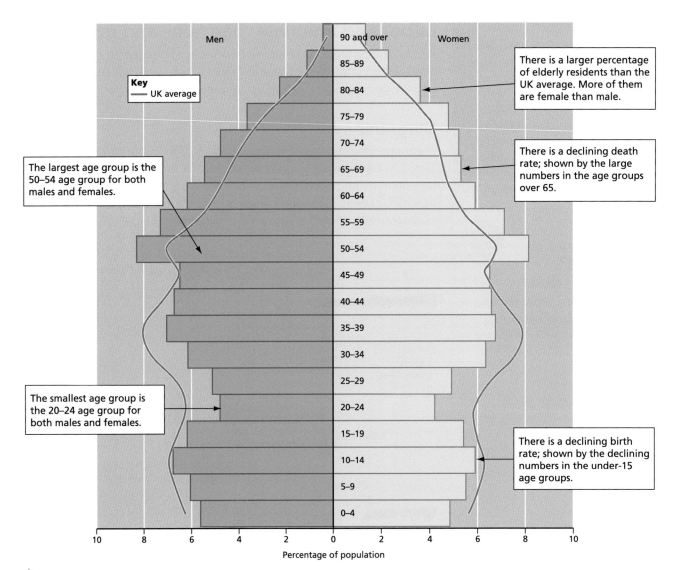

There is a larger percentage of elderly residents than the UK average. More of them are female than male.

The largest age group is the 50–54 age group for both males and females.

There is a declining death rate; shown by the large numbers in the age groups over 65.

The smallest age group is the 20–24 age group for both males and females.

There is a declining birth rate; shown by the declining numbers in the under-15 age groups.

Key
—— UK average

Percentage of population

 Figure 28 Population pyramid for Cornwall and the Scilly Isles

Exam Tip

If asked to annotate a graph or diagram, remember to give a detailed description or an explanation, not a simple descriptive point.

REVIEW

By the end of this section you should be able to:

✔ draw, label and annotate diagrams and graphs.

ACTIVITIES

1 Draw a population pyramid for the UK in 2014. Use the figures in the table (the figures have been rounded up or down).
2 Use annotations to describe and explain the shape of the graph.
3 Suggest reasons why these particular age groups have been used to display the age structure of the UK.
4 Work out the percentage of the population which is under 15.

Age	Males (m)	Females (m)
0–14	5.6	5.3
15–24	4.1	3.9
25–54	13.2	12.8
55–64	3.6	3.7
65 years and over	4.9	6.1

Graphical skills

Graphical skills you will:

→ study bar charts, histograms, divided bar charts and population pyramid charts
→ study line graphs, compound line graphs, flow lines, desire lines and isolines
→ study pie charts
→ study pictograms
→ study choropleth maps
→ study proportional symbols
→ study scatter graphs
→ study dot maps
→ study dispersion graphs.

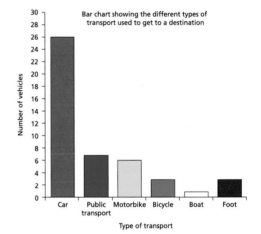

Bar chart showing the different types of transport used to get to a destination

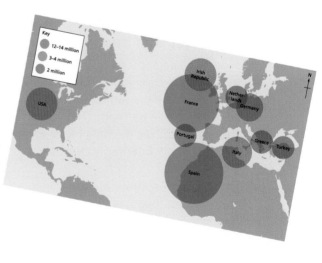

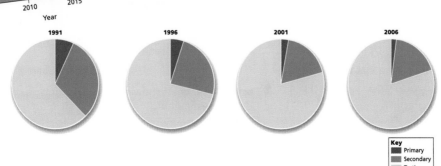

2 Graphical skills

⭐ Learning objective

– To study bar charts, histograms, divided bar charts and population pyramid charts.

⭐ Learning outcomes

→ To be able to construct each type of graph.
→ To be able to describe and explain the patterns that the graphs show.
→ To be able to suggest appropriate uses of these types of graphs.

Exam Tip

Remember that bar charts show discrete data; histograms show continuous data.

Different types of bar graphs

What is the difference between a bar chart and a histogram?

There are a number of differences between bar charts and histograms that students need to be aware of. Bar charts are one of the simplest forms of displaying data. Each bar is the same width but of varying length, depending on the figure being plotted. The bars should be drawn an equal distance apart. The data used for a bar chart is discrete data. This means that it is a discrete piece of information. An example would be the different types of vehicles in a traffic count.

When drawing a histogram, the bars should be drawn touching each other because a histogram is used to portray continuous data. An example would be the traffic flow for a continuous time frame.

How to draw a bar chart for the different types of transport used to get to a destination

- Decide on an appropriate scale on the x-axis for the bars. Remember that the bars should be the same width with a gap in between each one. This can be the same width as the data bar or a different width but the width must be a regular size.
- Decide on an appropriate scale on the y-axis for the number of vehicles. Remember that the scale should be spaced out evenly and allow for the highest value in the data set.
- Draw each of the bars to the correct value.
- The bars should be coloured in different colours as the data is discrete.
- Don't forget to label your axes and put a title on your graph.

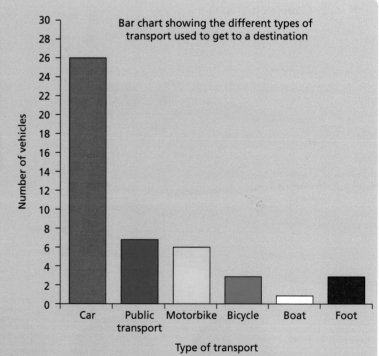

⬆ **Figure 1** Bar chart showing the different types of transport used to get to a destination

How to draw a histogram with equal class intervals of pedestrian flow for a continuous time scale

- Decide on an appropriate scale on the x-axis for the bars. Remember that the bars should be the same width with no gap in between.
- Decide on an appropriate scale on the y-axis for the number of people. Remember that the scale should be spaced out evenly and allow for the highest value in the data set.

- Draw each of the bars to the correct value.
- The bars should be coloured in the same colour as the data is continuous.
- Don't forget to label your axes and put a title on your graph.

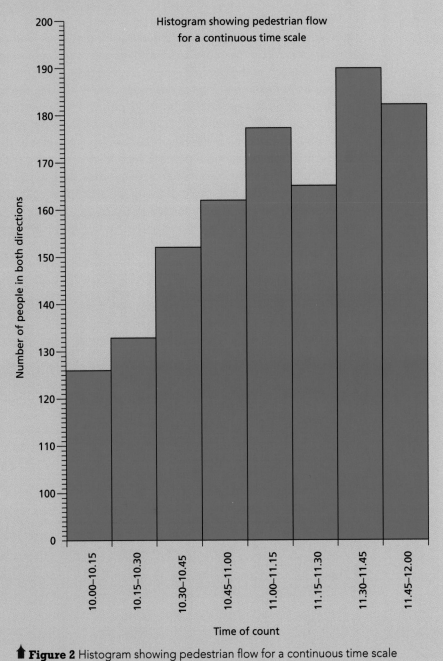

Figure 2 Histogram showing pedestrian flow for a continuous time scale

What is a divided bar chart?

In divided bar charts, the bars are subdivided on the basis of the information being displayed – in this example the sectors of industry in different countries. Each bar is out of 100 per cent.

How to draw a compound (divided) bar chart for sectors of industry in certain countries

■ These can be drawn with the bars joined or an equal distance apart. In this example the bars are joined together.

■ Decide on an appropriate scale on the x-axis for the per cent employed. Remember that the scale should be spaced out evenly.

■ Decide on an appropriate width of bar for the y-axis. Remember that the bars should be the same width with no gap in between them.

■ Divide each of the bars into the correct percentage for the industrial sector. For example, France has 6 per cent in primary, the line will be drawn at 6 per cent on the graph; 28 per cent in secondary, the line will be drawn at 34 per cent because it builds on the 6 per cent which is already there; tertiary has 66 per cent which is the rest of the space.

■ Each sector, for example primary, secondary or tertiary, should be coloured the same colour in each of the bars.

■ When complete, the whole of the graph should be coloured in. In this example, there will be three colours used.

■ Don't forget to label your axes and put a title on your graph.

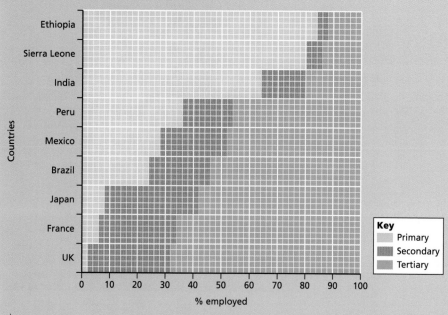

⬆ **Figure 3** A compound (divided) bar chart showing the sectors of industry in certain countries

What is a pyramid chart?

Another form of histogram is the pyramid chart. This is usually used to portray population (age–sex) data. It is constructed in a number of ways, but usually as 5- or 10-year age groups, with males on one side and females on the other. The lines are drawn horizontally and are the same width. The length of the bars is determined by the number of people in that age group or the percentage of the total population that age group represents. This type of chart can be used for any continuous data, for example a pedestrian count for a continuous time scale for movement in two directions (see Figure 2).

How to draw a pyramid chart for a pedestrian count

- Use the graph paper in landscape view.
- Either draw two lines vertically up the centre of the paper a small distance apart, or draw one line as shown in Figure 4.
- The x-axis is now divided into two halves. The left-hand side of the graph paper is for the flow in one direction, the right-hand side is for the flow in the other direction. Decide on an appropriate scale for the x-axis. It should be the same for each side of the graph. Remember that the scale should be spaced out evenly and allow for the highest value in the data set. Decide on an appropriate width of bar for the y-axis. Remember that the bars should be the same width with no gap in between them.
- Draw each of the bars to the correct value.
- The bars should be coloured in the same colour as the data is continuous.
- Don't forget to label your axes and put a title on your graph.

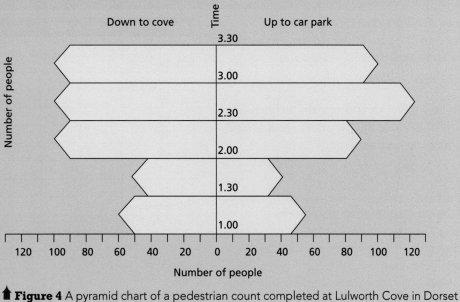

▲ **Figure 4** A pyramid chart of a pedestrian count completed at Lulworth Cove in Dorset

What is a located bar chart?

Bar charts can be placed on to maps to give the data more relevance. In this case, the bar charts have been drawn at the appropriate place on the map. Another way of doing this would be to draw the bar charts on tracing paper and use the tracing paper as an overlay. It is also possible to draw the bar charts on graph paper and paste them on to the map.

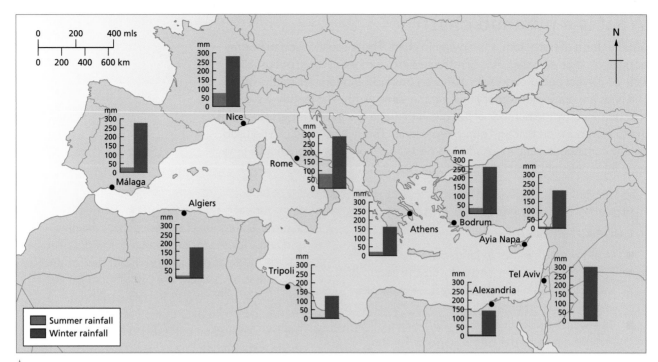

⬆ **Figure 5** Seasonal water supply in the Mediterranean

ACTIVITIES

1 Draw and complete the population pyramid for Cambridge shown in Figure 6. Use the data in this table and the 'how to' box on page 37 to help you.

2 Describe the shape of the pyramid. Use data in your answer.

3 Give reasons why this is an appropriate technique for this type of data. Choose the two most appropriate statements from the list below.
   – It shows that there is a relationship between the data sets.
   – It clearly shows whether any of the age groups has more of the population than the others.
   – The patterns in the data cannot be clearly seen. The data is discrete.
   – The data is continuous.

4 State a graphical technique other than pyramids that could be used to display pedestrian flows.

5 Give a reason for your choice of technique in question 4.

6 Describe the pattern of seasonal water supply in the Mediterranean shown in Figure 5. Base your description on the answers to the questions below.
   – What was the highest precipitation in the summer and which country received it?
   – What was the highest precipitation in the winter and which country received it?

Age	Males X1000	Females X1000
0–9	5	4
10–19	7	6
20–29	15	12
30–39	8	7
40–49	6	6
50–59	5	4
60–69	3.5	3
70–79	3	4
80–89	1	2.5
90 and over	1	2

- Which country had the greatest difference between summer and winter totals?
- What was the amount?
- Which country had the smallest difference between summer and winter totals?
- What was the amount?

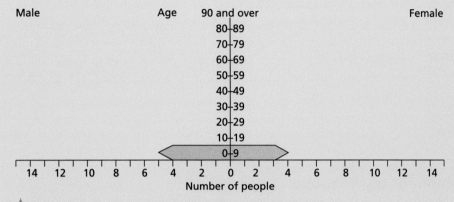

Figure 6 Part of the population pyramid for Cambridge

STRETCH AND CHALLENGE

For the graphical technique you chose in question 4, use the data for pedestrian counts provided in Figure 4. Draw your suggested technique. Annotate it with the reasons why you think it is an appropriate technique.

Exam Tip

Remember, on the exam paper you can be asked why a particular graphical technique is appropriate.

REVIEW

By the end of this section you should be able to:

✔ construct bar charts, histograms, compound bar graphs and pyramid graphs
✔ describe and explain the patterns that the graphs show
✔ use the graphs appropriately.

Line graphs, compound line graphs, flow lines, desire lines and isolines

What is a line graph?

A line graph is a basic graphical technique used to show changes over time (continuous data). In all line graphs, there are independent and dependent variables. Line graphs can be used to show multiple sets of data, for example traffic counts for the same place at different times.

How to draw a line graph for traffic counts

- The times of the count are the independent variable because the counts are dependent on when they are completed. These should be plotted on the horizontal x-axis.
- The number of vehicles is the dependent variable and should be plotted vertically on the y-axis.
- Decide on a scale appropriate for the range in values to be plotted on the x- and y-axes.
- Plot each line in turn. There should be four lines on your completed graph.
- They should be drawn in different colours and a key provided.
- Don't forget to label your axes and put a title on your graph.

| Times | Friday 6 May | | Sunday 8 May | |
	Into Lulworth	Out of Lulworth	Into Lulworth	Out of Lulworth
12.30–1.00	30	41	117	71
1.00–1.30	42	44	113	76
1.30–2.00	27	41	123	45
2.00–2.30	30	32	136	53
2.30–3.00	23	13	83	72
3.00–3.30	15	10	65	80

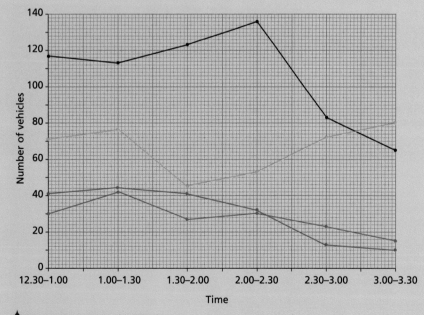

▲ **Figure 7** Graph of traffic flows into and out of Lulworth Cove

What is a compound line graph?

In compound line graphs, the graph is subdivided on the basis of the information being displayed in lines across the graph. For example, in Figure 8 the USA's percentage of world usage of renewable fuels has declined from 29 per cent in 1990 to a projected 20 per cent in 2030.

How to draw a compound line graph for renewable energy usage

- The years are the independent variable and should be plotted on the horizontal x-axis.
- The percentages are the dependent variable and should be plotted vertically on the y-axis.
- Decide on a scale appropriate for the range in values to be plotted on the x- and y-axes.
- Plot the data for the first country, the USA, on the graph.
- Add the percentage for the second country to the percentage for the first country. This should

be calculated for each year, for example USA in 1990 is 29 per cent; add 14 per cent for Canada. The point for Canada should be plotted at 43 per cent.
- The area between the countries should be coloured in a different colour for each section. Ideally, the order of the countries should be largest to smallest with colours that can be clearly identified.
- Don't forget to label your axes and put a title on your graph.

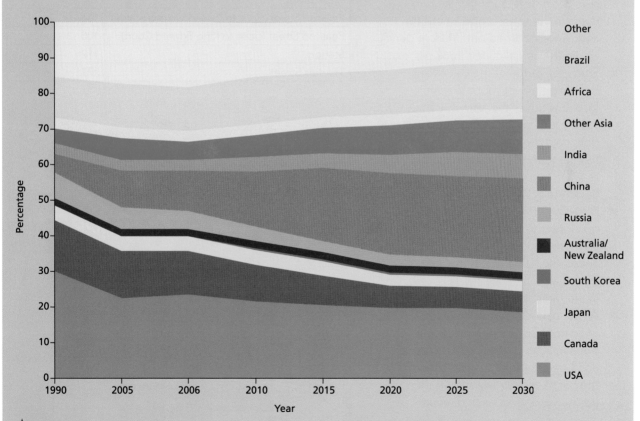

⬆ **Figure 8** World usage of renewable energy by region 1990–2030 (Source: US Energy Information Administration data)

What is a desire line?

A desire line is drawn to show the direction of movement. The line is drawn from the origin to the destination of a particular movement. For example, traffic flows from the suburbs of a city to the city centre.

What is the difference between a flow line and an isoline?

Flow line maps and isoline maps are more sophisticated ways of portraying data.

A flow line shows movement between places. The thickness of the line indicates the amount of movement. The direction can be shown by an arrow. Figure 9 shows pedestrian flows at a number of places around Windsor town centre. The arrows point in both directions, showing that the total is a combination of both directions.

How to draw a flow line map for pedestrian counts in Windsor

- Draw a base map which shows the relevant details, such as the main roads in the town centre where the counts were carried out. Mark in pencil where the actual counts took place.
- Study the range of the values and decide on an appropriate scale for the width of the arrows. If the scale is too large the flow lines will dominate the map, so take care at this stage to use the appropriate scale. In Figure 9 the scale is 5 mm thickness = 100 pedestrians.
- Draw the flow lines; these should go along the road where the count took place. The tail of the arrow should be where the flow began and the nose should point in the direction of the flow.
- As the count is a combination of both directions, the arrow will point in both directions. If a note had been made of the direction of the count, two arrows should then be drawn, one for each direction.

Street	Pedestrian flow
William Street	210
St Leonards Road	112
High Street	159
Peascod Street (close to Queen Victoria Statue)	166
Peascod Street (close to King Edward Court)	100
Station	112
Thames Street	277

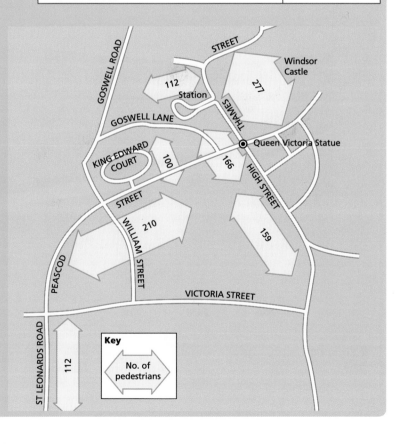

➡ **Figure 9** Pedestrian flows in Windsor

An isoline joins places of equal value and shows the distribution of an item over a particular area. Figure 10 shows places that are equal walking time from a shopping centre; these are known as isochrones. The most well-known isolines are:

- contours which join places on a map of equal height
- isobars which join places of equal pressure on a weather map
- isovels which join places of equal velocity in a river.

How to draw an isoline map for time distance from a retail centre

- Plot the data onto a map of the area. This should be a series of points which have a fairly even spatial distribution.
- Decide on the interval that you want between the isolines. If this is too small, then the map will appear cluttered. If it is too great, the map will become too generalised.
- Draw in your isolines, in this case every 5 minutes.
- The space between the isolines can be left blank or shaded in. If shaded, the colour should become greater as the value becomes higher.
- If you shade your map, then you should provide a key.

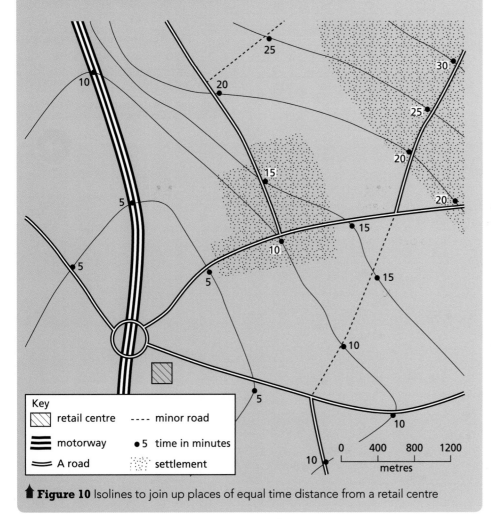

Key

- ▨ retail centre
- ≡ motorway
- ⌇ A road
- - - - minor road
- •5 time in minutes
- ⋰⋱ settlement

0 400 800 1200
metres

⬆ **Figure 10** Isolines to join up places of equal time distance from a retail centre

ACTIVITIES

1 Study the line graph for traffic flows in and out of Lulworth Cove on page 40. Suggest two advantages and two disadvantages of this display technique.

2 A line graph is one way of portraying traffic count data. Choose another type of graphical technique.
   a Draw the graph.
   b Give reasons why this is an appropriate technique for this type of data.
   c Describe the pattern shown on the graph.

3 Study Figure 9 on page 42, which is a flow line map of pedestrian counts in Windsor. State four ways in which this map has been incorrectly drawn.

4 Redraw the map of pedestrian flows in Windsor correctly.

5 State another graphical technique that could be used to display this data.

6 Complete a compound line graph for the information in the table below. The table shows different countries' percentage share of the world's gross domestic product (GDP).

	2005	2010	2015	2020	2025	2030
USA	23	22	21	21	20	18
Europe	28	24	22	18	16	13
Japan	10	9	8	6	5	4
China	4	9	12	14	16	18
India	3	4	5	8	10	12
Rest of Asia	6	6	7	8	8	10
South America	6	6	6	6	6	6
Russia	4	4	3	3	3	3
Other	16	16	16	16	16	16

REVIEW

By the end of this section you should be able to:

✔ construct each type of graph
✔ describe and explain the patterns that the graphs show
✔ use the graphs appropriately.

STRETCH AND CHALLENGE

Contour lines are isolines. Trace the drawing and complete it by plotting the following contour lines: 140, 130, 110, 100, 90, 80 and 70. Use the 'how to…' box on page 43 to determine the type of slope shown on the diagram.

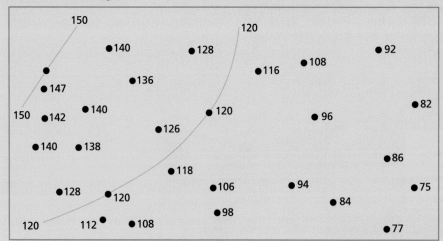

Pie charts

What is a pie chart?

A pie chart or divided circle is a basic graphical technique for showing a quantity which can be divided into parts. A pie chart can also be drawn as a proportional circle. (This is dealt with on pages 50–51.) Pie charts can be located on maps to show variations in the composition of a geographical phenomenon.

How to draw a pie chart for types of employment in rural areas

- If necessary, convert your data into percentages.
- You can either convert your percentages into degrees or use a percentage protractor to draw your pie diagram.
- Draw a circle and mark the centre.
- Subdivide the circles into sectors of the appropriate size.
- Differentiate the sectors by using different shadings or colours.
- Complete a key explaining the shadings and/or colours.
- Don't forget to put a title on your work.

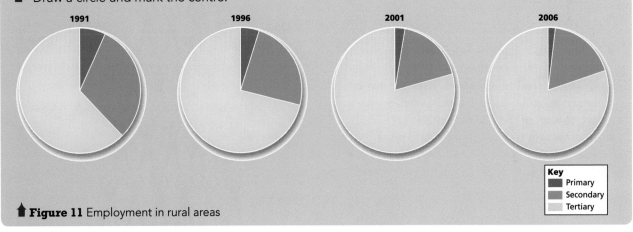

Key
- Primary
- Secondary
- Tertiary

 Figure 11 Employment in rural areas

ACTIVITIES

1 Construct a pie chart for the 2011 data shown in the table.

Type of employment in 2011	Percentage
Primary	2
Secondary	16
Tertiary	82

2 Outline the changes in employment shown between 1991 and 2006 in Figure 11.
3 Suggest another graphical technique that could be used to show this information.

STRETCH AND CHALLENGE

Complete a pie chart for the information below.

Justify your use of this technique.

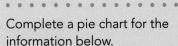

Type of water consumption	Percentage usage
Domestic	11
Agriculture	69
Industry	20

REVIEW

By the end of this section you should be able to:

✔ construct pie diagrams
✔ describe and explain the patterns that the diagrams show
✔ use the graphs appropriately.

Exam Tip

You may be required to construct, complete and interpret pie charts.

Pictograms

What is a pictogram?

A pictogram is a way of portraying data using appropriate symbols or diagrams that are drawn to scale. Pictograms can be a useful way of displaying data if accuracy is not so important and the data is discrete.

How to draw a pictogram for world population growth

■ Study the data that has to be displayed. Decide on the symbols to use to display the information. In this case the shape of a person could be drawn.
■ Decide the scale for the symbols. In this case 1 person = 1 billion people.
■ When constructing a pictogram, the years do not need to be continuous.
■ Draw the symbols. In some cases the symbol may not be full sized if it represents a proportion of the full amount.
■ Don't forget to provide a key and put a title on your graph.

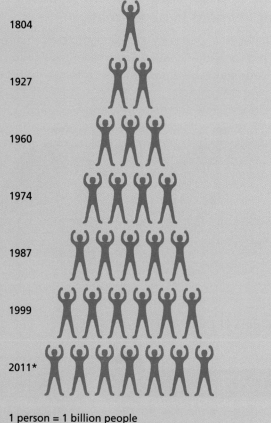

1804

1927

1960

1974

1987

1999

2011*

1 person = 1 billion people
* Estimated

⬆ **Figure 12** Pictogram for world population growth

ACTIVITIES

1 Construct a pictogram for the data below, which was taken from a questionnaire of how shoppers travelled to a shopping centre.

Mode of transport	Number of people
Car	26
Public transport	7
Taxi	1
Pedal bike	8
Motorbike	3
On foot	5

2 Choose another graphical technique. Display the data using this technique.
3 Suggest two advantages of using a pictogram.
4 Suggest two advantages of using the other technique you have chosen.

STRETCH AND CHALLENGE

Pictograms can be located on maps to produce a more sophisticated technique. Use the data below to draw pictograms on the map of Windsor (refer to the map on page 42 for the location of the streets).

Street	Pedestrian flow
William Street	210
St Leonards Road	112
High Street	159
Peascod Street (close to Queen Victoria Statue)	166
Peascod Street (close to King Edward Court)	100
Station	112
Thames Street	277

Exam Tip

When drawing pictograms, always remember to provide a key for the symbols.

REVIEW

By the end of this section you should be able to:

✔ construct pictograms
✔ describe and explain the patterns that pictograms show
✔ use pictograms appropriately.

Choropleth maps

What is a choropleth map?

A choropleth map is a map that is shaded according to a prearranged key, each shading or colour representing a range of values. The colours should become darker as the numbers increase. There are some inaccuracies in using this technique; one of these is that variations within units are concealed. It also gives a false impression of abrupt changes at boundaries. However, choropleth maps are easy to complete and show a good visual impression of change over space.

How to draw a choropleth map for average precipitation in the UK (see Figure 13)

- Obtain a base map of the UK.
- Find the range of your values and devise a shading scale.
- You should try to have no fewer than four shading bands and no more than eight.
- The shading should get darker as the value gets higher.
- Complete the map by shading in the areas.
- Draw a key for the map.
- Don't forget to put a title on your map.

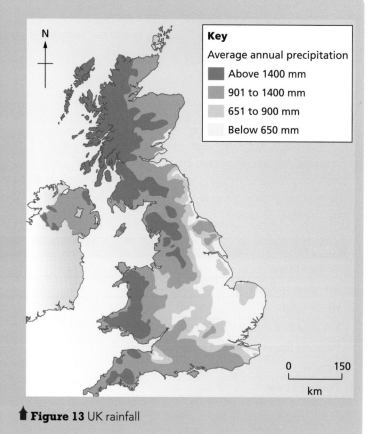

Key

Average annual precipitation

- Above 1400 mm
- 901 to 1400 mm
- 651 to 900 mm
- Below 650 mm

↑ Figure 13 UK rainfall

ACTIVITIES

1   Draw a choropleth map for the information on the top ten tourist destinations in 2009, shown in the table. Use the 'how to' box on page 48 to help you.

Destination	Tourist arrivals (millions)
France	81.9
Spain	59.2
USA	56.0
China	54.7
Italy	43.7
UK	30.7
Germany	24.4
Ukraine	23.1
Turkey	22.2
Mexico	21.4

2   Describe the pattern shown by the map.
3   A choropleth map is one way that this data could be displayed. Suggest two advantages and two disadvantages of using a choropleth map for this data.
4   Suggest and complete another sophisticated data presentation technique that could have been used to display the data.
5   State one way that this technique is more appropriate than a choropleth map and one way that it is less appropriate than a choropleth map.
6   A technique for displaying this data would be a histogram. Why would a histogram be drawn rather than a bar chart?

STRETCH AND CHALLENGE

Use the data in the table below to draw a choropleth map of the source countries of migrants to the UK from Eastern European countries between 2004 and 2006.

Country	Number of migrants (thousands)
Czech Republic	25
Estonia	10
Hungary	20
Latvia	28
Lithuania	55
Poland	260
Slovakia	40
Slovenia	5

Exam Tip

Questions could be set on completing a choropleth map and a key. Therefore, be sure that you understand the rules for a successful choropleth map.

REVIEW

By the end of this section you should be able to:

✔ draw choropleth maps
✔ describe and explain the patterns that choropleth maps show
✔ use choropleth maps appropriately.

Proportional symbols

What is a proportional symbol?

These are symbols that are drawn in proportion to the size of the variable being represented. The symbol could be anything, for example a pictogram could be used, but is more usually circles or squares.

How to draw proportional circles and squares on a map

- Find a base map of the area for the values you are displaying.
- Calculate the square root of each of the values (see table opposite). To work out the square root for Spain, put 14 into a calculator and press the square-root key. The answer should be rounded to one decimal place, so 3.7 is accurate enough.
- This answer will be used to construct your proportional symbol, either the radius of a circle or the side of a square.
- Study the square-root values to determine the range.

- Determine a scale that suits the range of square-root values. Remember that the symbol must fit on the base map. For example, in Figure 14 a scale of 5 mm is used. Multiply each square-root value by 5 to determine the symbol size. Therefore, the proportional symbol for Spain has a radius of 18 mm.
- Locate the points on the map where you are going to draw your symbol.
- Draw the symbols and shade or colour them.
- Provide a key for the scale and put a title on your map.

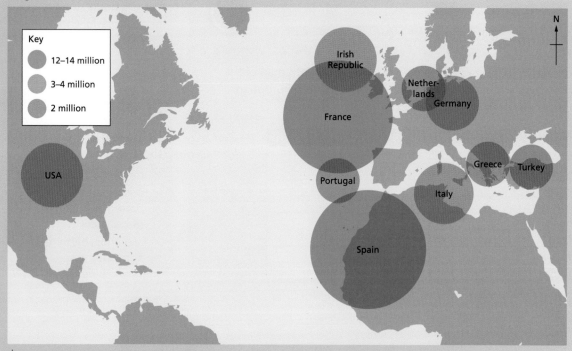

Figure 14 Proportional circles showing the number of British tourist visits to overseas tourist destinations

Key
- 12–14 million
- 3–4 million
- 2 million

Top ten overseas tourist destinations	Number of British tourist visits (rounded to the nearest million)
Spain	14.0
France	12.0
USA	4.0
Irish Republic	4.0
Italy	3.5
Germany	3.0
Portugal	2.0
Greece	2.0
Netherlands	2.0
Turkey	2.0

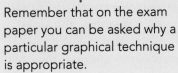

Exam Tip

Remember that on the exam paper you can be asked why a particular graphical technique is appropriate.

ACTIVITIES

1 Complete your own map of the top ten tourist destinations above by following the 'how to' instructions opposite.
2 Describe the pattern shown by the map.
3 State two advantages and two disadvantages of this technique.

STRETCH AND CHALLENGE

Complete a proportional circles map for the data in the table. You should draw pie charts for the information about water usage. The circles should be proportional to the gross domestic product in US dollars (GDP $) with the data within them.

Country	Domestic water usage (%)	Agriculture water usage (%)	Industry water usage (%)	GDP ($)
China	12	65	23	5,000
Australia	15	75	10	29,000
Japan	20	62	18	30,000
India	6	89	5	2,900
Russia	19	18	63	9,000
Uzbekistan	5	93	2	1,700
Malaysia	17	62	21	8,500
Sri Lanka	4	93	3	3,700
Afghanistan	3	96	1	700
USA	17	42	41	37,800

REVIEW

By the end of this section you should be able to:

✔ construct proportional symbols
✔ describe and explain the patterns that proportional symbols show
✔ use the graphs appropriately.

Scatter graphs

What is a scatter graph?

A scatter graph can be used to show whether there is a relationship (link) between two sets of data. The pattern of the points describes the relationship. After plotting the points, a line known as a best-fit line should be drawn on the graph. This line will indicate the strength of the relationship (correlation) between the two variables (data sets). The pattern will show a positive or negative correlation or no correlation at all. Study the graphs in Figure 16, which show scatter graphs with different correlations.

How to draw a scatter graph to show whether there is a graphical correlation between the width and depth of a river as it moves from its source (site 1) towards its mouth (site 10)

■ Decide which is the independent variable and which is the dependent variable. For these two sets of data, there are no independent or dependent variables. However, if you were plotting how depth changes with distance from the source, the distance from the source would be the independent variable and the depth would be the dependent variable.

■ Decide on an appropriate scale on the x-axis for the width measurements. Remember, the scale should be spaced out evenly and allow for the highest value in the data set. In this case, ten squares on the graph paper equals 1 metre.

■ Decide on an appropriate scale on the y-axis for the depth measurements. Remember, the scale should be spaced out evenly and allow for the highest value in the data set. In this case, five squares on the graph paper equals 10 cm.

■ Plot the measurements for each of the sites onto the graph, labelling each site with the correct number.

■ Draw a line of best-fit. This is a straight line through the middle of the points that you have plotted.

■ Compare the pattern with the standard patterns for the different types of correlations shown in Figure 16.

■ What type of correlation have you plotted?

■ Explain what this means.

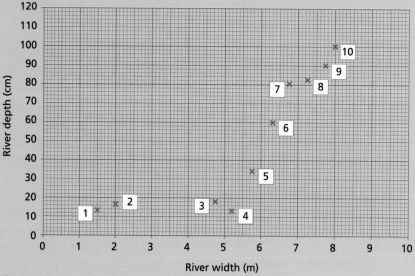

▲ **Figure 15** A scatter graph of river width and river depth

ACTIVITIES

1 Draw a scatter graph for the data below. Use the 'how to' box opposite to help you.

Country	Domestic water usage (%)	GDP ($)
China	12	5,000
Australia	15	29,000
Japan	20	30,000
Thailand	9	7,400
Korea	14	17,700
India	6	2,900
Indonesia	8	3,200
Russia	19	9,000
Turkey	15	6,700
New Zealand	48	21,600
Uzbekistan	5	1,700
Malaysia	17	8,500
Sri Lanka	4	3,700
Algeria	25	5,900
Aghanistan	3	700
Sierra Leone	2	500
USA	17	37,800

2 Is there a correlation? What is its nature?
3 Points that are well away from the line of best-fit are known as residuals or anomalies. Are there any residuals or anomalies? If so, circle them on your graph.
4 Describe and give reasons for the pattern that is shown by the graph.

STRETCH AND CHALLENGE

Another way to test for a relationship between sets of data is to use a statistical technique such as Spearman's rank correlation coefficient. Test the statistical correlation between the data given in question 1 using Spearman's test. Information on how to complete this statistical technique can be found in the Stretch and Challenge section in Chapter 3 on page 64.

REVIEW

By the end of this section you should be able to:

✔ construct a scatter graph
✔ explain the patterns shown on a scatter graph
✔ suggest appropriate uses of scatter graphs.

Exam Tip

Remember to always state the type of correlation and explain what it means.

Positive correlation

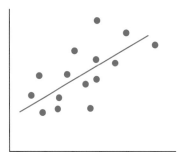

The line of best-fit stretches from the bottom left to the top right of the graph. This indicates a positive correlation; as one variable increases so does the other variable.

Negative correlation

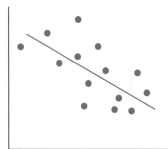

The line of best-fit stretches from the top left to the bottom right of the graph. This indicates a negative correlation; as one variable increases the other variable decreases.

No correlation

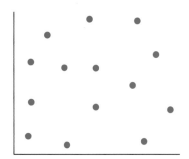

The points are distributed all over the graph. This shows that there is no relationship between the variables.

Figure 16 Scatter graphs with different correlations

Dot maps

What is a dot map?

Dot maps are drawn with dots of a fixed size which represent a given value such as numbers of people. The position of the dot on the map shows the location on the map of that variable. They are very useful for showing distributions where the values are a fixed amount and the location of the variable is known. However, if there are a large number of dots they can be very difficult to count and the maps difficult to construct.

How to draw a dot map of cities with at least 10 million inhabitants

■ Find a map of the world and the information you require on cities with at least 10 million inhabitants.
■ Decide on the size of the dot. You already have the value of the dot, which will be at least 10 million inhabitants. Dots should be a uniform size that is appropriate for the size of the map and the density of the data being displayed.

■ Locate the points on the map where you are going to place your dot using an atlas.
■ Draw the dot on the map in the appropriate place with a number and provide a key. You may wish to write the name of the city next to the dot but a number with a key would be better. Ensure that the key box and the map have a title.

⬆ **Figure 17** Some of the world's cities with at least 10 million inhabitants in 2015

ACTIVITIES

1   Study Figure 17. Describe the distribution of the world's cities with at least 10 million inhabitants shown on the map.
2   Explain the advantages and disadvantages of using this mapping technique.
3   Suggest another technique that could be used to display this information.

STRETCH AND CHALLENGE

Research information on ethnicity in a city. Produce a dot map to show the distribution of the members of the five largest ethnic groups. Describe the distribution that is shown.

Exam Tip

When drawing a dot map, ensure that you choose an appropriate value for the dot size so that the dots do not appear crowded on the map.

REVIEW

By the end of this section you should be able to:

✔ construct a dot map
✔ explain the patterns shown on a dot map
✔ suggest appropriate uses of these types of dot maps.

Dispersion graphs

What is a dispersion graph?

A dispersion graph shows the range of a set of data. It shows whether the data tends to group or disperse. It can also be used to compare sets of data. The values are plotted on the vertical axis. There is also a short horizontal axis which can show the frequency (the number of times) the variable occurs.

How to draw a dispersion graph for the size of pebbles found on a beach

- The study was completed at three sites, therefore the graph must allow for this.
- Study the values of the pebble sizes in the sample. Decide on a scale for the range of values.
- Plot the scale on the vertical axis.
- The horizontal scale is for the sites and if necessary allows you to repeat a value.
- Don't forget to label your axes and put a title on your graph.
- The mean size has also been marked on Figure 18 using blue diamonds.

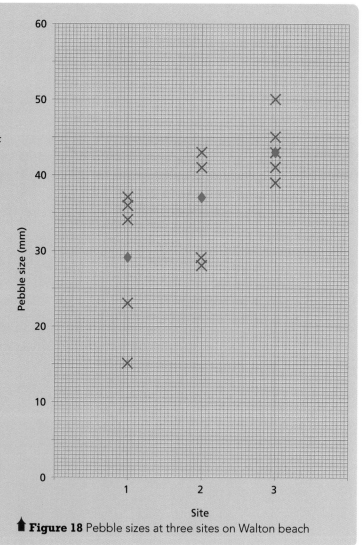

▲ **Figure 18** Pebble sizes at three sites on Walton beach

Powers' roundness index

Maurice Powers published a paper in 1953 describing a technique of measuring pebbles. This technique gives a visual representation against which the shape of pebbles can be compared. The chart, now called the Powers' roundness index, is split into six categories, each denoting a degree of roundness from very angular to well rounded. One disadvantage is that one person's opinion on the roundness or angularity of a pebble may differ from another's.

① Very angular

② Angular

③ Sub-angular

④ Sub-rounded

⑤ Rounded

⑥ Well rounded

⬆ **Figure 19** Powers' roundness index

STRETCH AND CHALLENGE

Dispersion diagrams can be used to determine the range of data by plotting the inter-quartile range. Plot the median, upper and lower inter-quartile values and the inter-quartile range on the dispersion diagram for river data.

ACTIVITIES

1 Using the information in the 'how to' box opposite, complete a dispersion graph for the data in the tables below. The tables show data on pebble sizes gathered by a group of students from two sites on the Afon Tarell in South Wales.

Site 1

10	5	10	5	12	13	9	12	11	8
10	13	13	14	12.5	13.5	13	15	7	9

Site 2

3	4	6	6	6	5	10	4	7	6
4	6	7	5	3	4	9	3	4	3

2 Outline the differences between the two samples.
3 Describe the pattern shown by the data.
4 Identify another graphical technique that could have been used to display this data.
5 Give reasons for your choice of technique.
6 The students also gathered information on the shape of the pebbles using Powers' roundness index (which gives a value to the shape of pebbles). How could they have displayed this information in the same graph?

REVIEW

By the end of this section you should be able to:

✔ construct dispersion graphs
✔ describe and explain the patterns that the graphs show
✔ use the graphs appropriately.

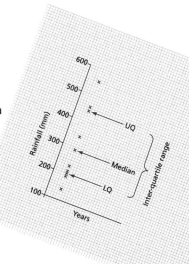

Numerical and statistical skills

In Chapter 3

Numerical and statistical skills you will:

→ study area, proportion, ratio, magnitude and frequency
→ work out the median, mean and mode of a set of data
→ work out the range and inter-quartile range of a set of data
→ calculate percentage increase and decrease for a set of data
→ calculate Spearman's rank correlation coefficient.

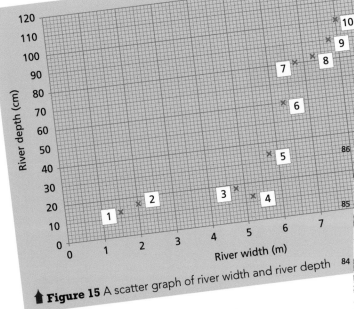

▲ **Figure 15** A scatter graph of river width and river depth

Arithmetic mean

$$\bar{x} = \frac{\sum x}{n}$$

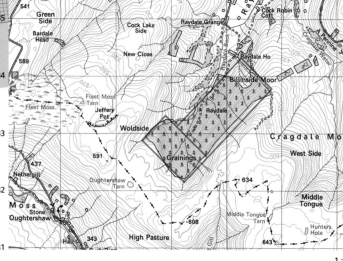

▲ **Note**: This map is not reproduced at original size.

1 :

3 Numerical and statistical skills

⭐ Learning objective

− To study numerical skills in geographical questions.

⭐ Learning outcomes

➜ To show an understanding of area.
➜ To understand and correctly use proportion and ratio, magnitude and frequency.

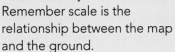

Exam Tip

Remember scale is the relationship between the map and the ground.

Numerical skills

Some numerical skills are discussed in other chapters of this book, such as scale in Chapter 1 and designing fieldwork data sheets, collecting data and drawing informed conclusions from numerical data which are dealt with in Chapter 4. This section deals with the elements of numerical skills which are not covered in other chapters.

Area

In geography, you could be asked about the area of part of a map or a photograph. An area is measured in square units. Therefore it is the length of an area times the width of an area.

For instance, if you want to know the area of a football pitch you will measure the length of the pitch and then the width of the pitch.

If the length of the football pitch is 100 m and the width of the football pitch is 50 m

100 x 50 = 5,000

So the area of the football pitch is 5,000 square metres (m^2).

Working out the area of a feature on a map can be a little more difficult because of the irregular space it can cover. Also, you must remember to use the scale on the map to obtain the actual measurement on the ground.

Worked example

Look at the OS Wensleydale map extract on page 86. What is the area of the woodland in grid squares 8882, 8982, 8883, 8983 and 9083?

In order to achieve a measurement for this area, the woodland will have to be measured in two small sections as it is different widths and lengths. The measurements are an estimate of the size. See Figure 1.

Box shown on Figure 1	Map measurement cm	Ground measurement km	Totals km
1	2 cm x 1 cm	1 km x 0.5 km	0.5
2	3 cm x 1.5 cm	1.5 km x 0.75 km	1.12

Therefore, 0.5 + 1.12 = 1.62 sq km is the area of the woodland.

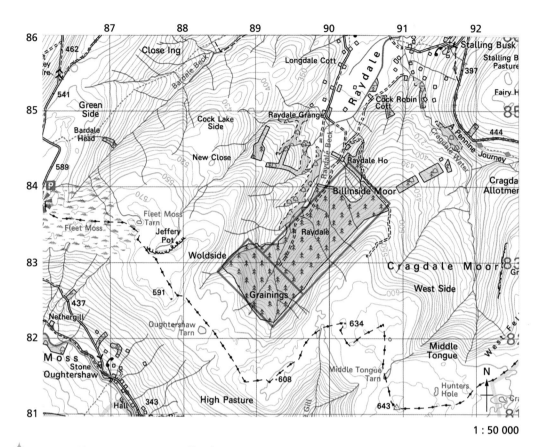

Figure 1 How to measure woodland on a map extract 1:50000

ACTIVITY

What is the area of Alnwick shown on the map extract on page 84?

Ratio, proportion, magnitude and frequency

A ratio is a comparison between two different things. In geography the most frequent ratio you will experience is the scale on maps, for example, 1:50000. This means that for every 1 cm on the map there are 50,000 cm on the ground. You could come across questions about ratios on the examination paper.

Proportions can be built from ratios, for example a half is a proportion of 5:10. For example, half of the forest could be deciduous trees and half coniferous trees.

In geography, magnitude is the relative size of something. The term is used most frequently when assessing the impact of an earthquake. It is a measurement of the energy released by the earth. However, the term could also be used in relation to settlements. One settlement could have double the population numbers of another. One area could have triple the amount of average rainfall of another area.

Frequency is how often something occurs. Therefore you could be asked about the frequency of volcanic eruptions or earthquakes in an area.

REVIEW

By the end of this section you should be able to:

✓ show an understanding of area

✓ understand and correctly use proportion and ratio, magnitude and frequency.

Statistical skills

How to work out the median, mean and mode of a set of data

Median, mean and mode are known as measures of central tendency. They are useful ways to analyse any set of geographical data.

Median

This is the middle value of a set of data. The numbers should be arranged in rank order. If the number of values is odd, then the median value will be:

$$\frac{n+1}{2}$$

For example, if the total number of values in a data set was 39, an odd number, the median would be the 20th value in the rank order of the data.

If there are an even set of values, the median is the mean of the middle two values.

Arithmetic mean

The (arithmetic) mean is calculated by adding up all of the values in the data set and dividing the total by the number of values in the data set:

$$\bar{x} = \frac{\sum x}{n}$$

Where:

$\sum$ = the sum of

$\bar{x}$ = arithmetic mean

n = number of variables

Exam Tip

Rank order means in order of size from smallest to largest or largest to smallest.

Mode

The mode is the value that occurs most frequently in a set of data. If the set of data is very large, it may be a good idea to group the data into classes before working out the mode. The class with the highest frequency, with the most values in it, is the modal class. The modal class does not have to contain the mode of the set of data.

How to calculate the range of a set of data

The range is the difference between the highest and lowest value in the data set. For example, when looking at temperatures for a settlement for the year, the range of temperature would be the difference between the highest and lowest

temperature. Therefore, using the figures in the table for London below:

Highest temperature = 19°C

Lowest temperature = 4°C

Annual temperature range = 19°C – 4°C = 15°C

London average monthly temperatures:

Month	Jan	Feb	Mar	April	May	June	July	Aug	Sept	Oct	Nov	Dec
Temp °C	4	5	8	10	12	15	18	19	15	10	7	4

How to work out the inter-quartile range of a set of data

The inter-quartile range (IQR) shows the spread of a set of data, but is more accurate than using the range because it doesn't include the extremities. The IQR indicates the spread of the middle 50 per cent of the data set, as it omits the top and bottom 25 per cent. It gives a better idea of how the data is spread around the median value. This can also be referred to as percentiles with the 50th percentile being 50 per cent of the data set, the 25th percentile being 25 per cent and so on.

The IQR is calculated by putting the data into rank order highest to lowest or plotting it on a graph. The values are then divided into four equal groups or quartiles, where n is the number of values.

The upper quartile (UQ) value is the value that occurs at the following position in the data set:

$$\frac{(n + 1)}{4}$$

The lower quartile (LQ) value is the value that occurs at the following position in the data set:

$$\frac{3(n + 1)}{4}$$

The difference between these two values is known as the IQR.

Worked example

Consider the rainfall data for a city in Israel shown in Figure 2. This data could be either ranked or drawn as a graph to work out the IQR.

Year	Average rainfall (mm)	Rank
2008	200	9
2006	140	11
2004	452	2
2002	190	10
2000	290	6
1998	442	3
1996	347	5
1994	560	1
1992	390	4
1990	209	8
1988	235	7

Figure 2 Rainfall data for a city in Israel

To work out the IQR for the rainfall data:

$$UQ = 11 + 1 = \frac{12}{4} = 3$$

Third position = 200

$$LQ = 3 \times 12 = \frac{36}{4} = 9$$

Therefore the IQR is 242 mm.

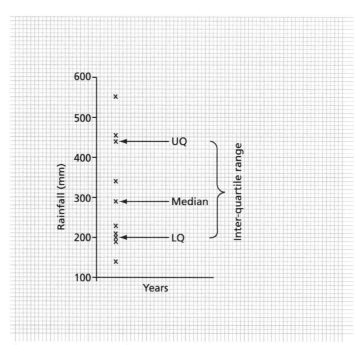

Figure 3 Dispersion graph of the rainfall data

ACTIVITIES

1 Work out the following measures of central tendency for the data in Figure 2.
   a Median
   b Mean

2 The results of a pebble survey along a beach in the south-east of England are shown in the table. The survey was completed using a quadrat. Ten pebbles were measured at each site.
   a Complete the table by working out the mean, mode and median for each site.

Sites	Distance along beach (m)	Size of pebbles (cm)	Mean pebble size (cm)	Mode pebble size (cm)	Median pebble size (cm)
1	0	8.5, 8, 7.5, 8, 6.5, 10.5, 4.5, 2, 7.5, 11			
2	50	8.5, 8, 7.5, 5, 6.5, 10.5, 4.5, 6, 7.5, 11			
3	100	6, 8, 7.5, 5, 6.5, 9.5, 4.5, 6, 6, 3			
4	150	8.5, 8, 7.5, 5, 6.5, 9.5, 4.5, 6.5, 6, 3			
5	200	8.5, 8, 7.5, 5, 6.5, 9.5, 4.5, 6, 6, 3			
6	250	4.5, 4, 7.5, 5, 6.5, 4.5, 4.5, 2.5, 3.5, 3			
7	300	4.5, 4, 3, 2, 3, 3.5, 5, 2.5, 3.5, 3			
8	350	4.5, 4, 2, 2, 8, 3.5, 2, 2.5, 3, 3.5			

   b Identify the advantages and disadvantages in each of these measures of central tendency.

How to calculate percentage increase or decrease in a set of data

There could be questions on the examination paper which ask you to calculate a percentage increase or decrease between two numbers. The way to work this out is to:

- work out the difference between the two numbers
- divide the increase (difference) by the first number
- multiply the answer by 100.

If your answer is a positive number, there has been a percentage increase between the numbers. If the answer is a negative number, then there has been a percentage decrease between the two numbers.

Worked example

For example, calculate the percentage increase in GDP for the UK.

In 2000 the UK had a GDP of US $1.5 billion. In 2014 it had a GDP of US$ 3 billion.

Therefore:

3 − 1.5 = 1.5 divided by 1.5 × 100 = 100%

The UK GDP between 2000 and 2014 had a percentage increase of 100%.

Country	GDP 2000 billions US$	GDP 2014 billions US$
Australia	0.4	1.4
China	1.2	10.3
India	0.4	2.0
USA	10.2	17.4
UK	1.5	3.0

Bivariate data

In Chapter 2, there is a section on scatter graphs (see page 52). There is information there on how to draw scatter graphs for bivariate data, how to sketch trend lines and lines of best-fit and how to describe the relationships that are shown on scatter graphs. However, you will also need to know how to interpolate and extrapolate trends.

- Interpolation is when a value is found inside the set of data.
- Extrapolation is when the value is found outside of the set of data.

ACTIVITIES

1　Study the graph on page 52 in Chapter 2. Interpolate the width for a depth of 65 cm. Extrapolate the value of the width for a depth of 110 cm.

2　Using the table on p. 63, calculate the percentage increase in GDP for Australia between 2000 and 2014.

REVIEW

By the end of this section you should be able to:

✓ use appropriate measures of central tendency – median, mean, range, quartiles and inter-quartile range, mode and modal class

✓ calculate percentage increase or decrease and understand the use of percentiles

✓ describe relationships in bivariate data: sketch trend lines through scatter plots, draw estimated lines of best-fit, make predictions, interpolate and extrapolate trends

✓ identify weaknesses in selective statistical presentation of data.

STRETCH AND CHALLENGE

Spearman's rank correlation coefficient

Spearman's rank correlation coefficient is a technique that is used to see whether there is a link between sets of data.

Here is a worked example to find out whether there is a correlation between settlement size and number of services in a settlement.

Step 1. The data should be put into a table, as shown, with the smallest number in the first set of data appearing first in column A, with its corresponding value in column C.

A Settlement population	B Rank	C Number of services	D Rank	E Differences between ranks (d)	F d²
190	12	4	11	1	1
330	11	3	12	1	1
1,550	10	11	10	0	0
2,660	9	19	9	0	0
5,632	8	39	8	0	0
8,890	7	41	7	0	0
9,950	6	87	4	2	4
10,500	5	43	6	1	1
10,900	4	92	3	1	1
11,890	3	75	5	2	4
14,330	2	109	2	0	0
16,789	1	144	1	0	0

Step 5. The differences should then be totalled. In this example this is a result of 12. So $\Sigma d^2 = 12$. You will need this in the next part of the calculation.

Step 4. The difference between the rankings, column E, is then squared and added to column F.

Step 2. Both sets of data should be given a rank order, with the largest observation given 1, as is the case for 16,789. This settlement also has the largest number of services, so it also has a rank of 1. Ranking should be placed in columns B and D.

Step 3. Work out the difference between the ranks of each variable and add to column E.

Step 6. Calculate the coefficient (r_s) using the formula: $r_s = 1 - \left(\dfrac{6 \Sigma d^2}{n^3 - n} \right)$

where Σ is the symbol for summation, d is the difference in rank of the values of each matched pair and n is the number of pairs.

Remember that $d^2 = 12$ from the above calculation and $n = 12$ (the number of pairs of data). Therefore:

$$1 - \left(\frac{6 \times 12}{12^3}\right) = 1 - \left(\frac{72}{1728 - 12}\right) = 1 - \left(\frac{72}{1716}\right) = 1 - 0.04 = 0.96$$

The result can be interpreted from the scale:

+1.0	0	−1.0
Perfect positive correlation	No correlation	Perfect negative correlation

In the above example a result of +0.96 indicates that there is a strong positive correlation between settlement size and the number of services at these sites.

Further development of the technique

You should now determine whether the correlation has happened by chance or not. To do this you must decide on the rejection level (a), this is, if you like, how confident you want to be about your findings. Therefore, if you wish to be 95 per cent certain, your rejection level is calculated as follows:

$$a = \frac{100 - 95}{100} = 0.05$$

Calculate the formula for t:

$$t = r_s \sqrt{\left(\frac{n - 2}{1 - r_s^2}\right)}$$

Where r is Spearman's rank correlation coefficient (= 0.96), n is the number of pairs (= 12) and df is the degrees of freedom (this means how confident you are about your results). Therefore, with a rejection level of 0.05 or 95 per cent:

$$t = r_s \sqrt{\left(\frac{n - 2}{1 - r_s^2}\right)} + 0.96 \sqrt{\left(\frac{12 - 2}{1 - 0.96^2}\right)} = 10.73$$

(the critical value)

You should now work out the degrees of freedom:

$$df = n - 2$$

Where n = number of pairs = 12.

Therefore

$$df = (n - 2) = (12 - 2) = 10$$

Look this up in the t-table (Figure 4) using the degrees of freedom of 10 and a 0.05 rejection level. Therefore, the critical value is 10 and the t-value on the t-table is 2.23.

So the t-value is less than the critical value, which means there is a significant correlation between settlement size and the number of services found in them.

If the critical value is less than your t-value, then the correlation is significant at the level chosen. If the critical value is more than your t-value, then you cannot be certain that the correlation did not occur by chance. This could be because:

- your sample was too small to permit you to prove a correlation
- the relationship is not a good one to have chosen.

ACTIVITY

Work out if there is a correlation between the distance from a groyne on a beach and the size of shingle on the beach.

Distance from groyne (m)	Size of shingle (cm)
50	7.6
100	7.9
160	8.2
300	6.8
420	7.4
480	6.8
550	5.8
670	7.9
850	6.8
950	4.7
1050	5.6
1150	4.2

Remember, if you have a joint rank (that is when two numbers in the table have the same value) you should add them together and then divide by the number there are, missing out the next numbers until you have omitted how many you add together.

Degrees of freedom	Rejection level probabilities				
	p = 0.1	p = 0.05	p = 0.02	p = 0.01	p = 0.001
1	6.31	12.71	31.82	63.66	636.62
2	2.92	4.30	6.97	9.93	31.60
3	2.35	3.18	4.54	5.84	12.94
4	2.13	2.78	3.75	4.60	8.61
5	2.02	2.57	3.37	4.03	6.86
6	1.94	2.45	3.14	3.71	5.96
7	1.90	2.37	3.00	3.50	5.41
8	1.86	2.31	2.90	3.36	5.04
9	1.83	2.26	2.82	3.25	4.78
10	1.81	2.23	2.76	3.17	4.59
11	1.80	2.20	2.75	3.11	4.44
12	1.78	2.18	2.68	3.06	4.32
13	1.77	2.16	2.65	3.01	4.22
14	1.76	2.15	2.62	2.98	4.14
15	1.75	2.13	2.60	2.95	4.07
16	1.75	2.12	2.58	2.92	4.02
17	1.74	2.11	2.57	2.90	3.97
18	1.73	2.10	2.55	2.88	3.92
19	1.73	2.09	2.54	2.86	3.88
20	1.73	2.09	2.53	2.85	3.85
21	1.72	2.08	2.52	2.83	3.82
22	1.72	2.07	2.51	2.82	3.79
23	1.71	2.07	2.50	2.81	3.77
24	1.71	2.06	2.49	2.80	3.75
25	1.71	2.06	2.49	2.79	3.73
26	1.71	2.06	2.48	2.78	3.71
27	1.70	2.05	2.47	2.77	3.69
28	1.70	2.05	2.47	2.76	3.67
29	1.70	2.05	2.46	2.76	3.66
30	1.70	2.04	2.46	2.75	3.65
40	1.68	2.02	2.42	2.70	3.65
60	1.67	2.00	2.00	2.66	3.46

Figure 4 A t-table

Geographical
enquiry skills

4

In Chapter 4

Geographical enquiry skills you will:

→ identify questions and routes to enquiry
→ design fieldwork data collection sheets, and collect data with an understanding of accuracy
→ carry out risk assessments
→ use quantitative and qualitative data
→ write conclusions
→ develop an extended argument.

4 Geographical enquiry skills

⭐ Learning objective

— To study geographical enquiry, literacy and ICT skills.

⭐ Learning outcomes

→ To be able to identify questions and sequences of enquiry.
→ To be able to design fieldwork data collection sheets and collect data with an understanding of accuracy, sample size and procedures, control groups and reliability.
→ To understand the need for and be able to carry out risk assessments.
→ To be able to use qualitative and quantitative data from both primary and secondary sources to obtain, illustrate, communicate, interpret, analyse and evaluate geographical information.
→ To be able to draw well-evidenced and informed conclusions about geographical questions and issues.
→ To write descriptively, analytically and critically, displaying good literacy skills, and to be able to communicate ideas effectively for a range of target audiences.
→ To be able to develop an extended written argument.

Geographical enquiry, literacy and ICT skills

Geographical enquiry and ICT skills will be examined on Paper 3 of the examination but they could also appear in either of the other two papers. You will need to be able to show your understanding of the processes involved in carrying out a geographical enquiry. You will not be expected to have access to a computer in the exam, but may have to show your understanding of the principles and use of ICT in geographical enquiries. Questions could be asked on how to extract information from the internet or how to use databases such as the census.

Identify questions

All geography students should be able to identify geographical questions. For example, if you look at a particular location, the questions you should be asking yourself are:

- What is the landscape like?
- What are the features that stand out?
- Where is this place – grid references?
- What is the area around it like?
- Why is it like it is?
- What is happening to certain variables?

From asking geographical questions, you should be able to form hypotheses on which a piece of work could be focused. A hypothesis is a statement that can be tested. In an exam question, you might be asked to formulate hypotheses or questions from a sample of information.

A geographical issue is a debatable point. For example, should a wind farm be located in a particular place? Questions such as the following could then be asked:

- What will be the impact on the environment?
- What will be the impact on local people?
- Are the climatic conditions correct?

Hypotheses could be set up and tested, or questions could be set and answered, but it is unlikely that a clear answer will be determined because so many different viewpoints are involved. Issues usually require candidates to make judgements from inconclusive data or evidence.

Whether you are studying a geographical question, hypothesis or issue, you should be able to explain why you are studying it and the results you expect to find out.

You could be given this task question and asked to develop hypotheses or questions to help answer it.
How do channel characteristics vary along a river?

Sample hypotheses for the rivers task question

- The velocity of the river increases as it moves from its source to its mouth.
- The width of the river increases as it moves from its source to its mouth.
- The depth of the river increases as it moves from its source to its mouth.
- The gradient of the river channel increases as it moves from its source to its mouth.
- The wetted perimeter of the river increases as it moves from its source to its mouth.

Sample questions for the rivers task question

- How does the velocity of the river change as it moves from its source to its mouth?
- How does the width of the river change as it moves from its source to its mouth?
- How does the depth of the river change as it moves from its source to its mouth?
- How does the gradient of the river channel change as it moves from its source to its mouth?
- How does the wetted perimeter of the river change as it moves from its source to its mouth?

The next stage would be to explain why each of the hypotheses/questions are relevant. They may relate to a geographical theory you have been taught.

Establish sequences of enquiry

Exam Tip

It is a good idea to use hypotheses rather than questions because often hypotheses are easier to answer.

All geographers should be able to establish a sequence of enquiry. This is sometimes called a road map. What it means is planning out how you will complete your investigation. You must establish what you need to do to answer your task question and the techniques you will need to complete to derive the data you require. You also need to plan when you will be doing each stage of the enquiry and set yourself deadlines. When you write up your study, it should be in the same sequence with each section in a different chapter. There should be good linkage between the chapters, which shows the sequence of the enquiry.

You could be asked to plan the sequence of a geographical enquiry.

A brief sample road map for a river study

Planning stage:	location and hypotheses/questions
Data collection:	when should I go on my field trip? what techniques should I use?
Data presentation:	what techniques shall I use?
Analysis:	explaining my results
Conclusions:	answering my questions
Evaluation:	what was the value of my study? how successful were my techniques?

Fieldwork data collection

In the examination, you could be asked how you would collect data on a particular topic such as a river or coastal enquiry. You would have to be able to prepare data collection sheets or illustrate ones that you had used in the field. You would have to show awareness of an understanding of the importance of accuracy, sample size, control groups and reliability. You would also need to know about the use of spreadsheets and data-handling software in the field.

Data collection sheets

You can use programmes such as Word™ or Excel™ to produce data collection sheets. These can provide questionnaires and spreadsheets to enable data to be easily collected in the field using iPads or tablets, or on mastersheets using a clipboard.

Figure 1 shows an Environmental Quality Analysis (EQA) which has been prepared using a table in Word. It can be either completed on a mobile device such as a tablet, or printed and completed by hand. Data handling software on data loggers can be used in the field to collect vast amounts of information such as temperature, speed of river flow or simple traffic counts (see Figure 2). The information collected can then easily be transferred onto spreadsheets and displayed.

Criteria	+3	+2	+1	0	–1	–2	–3
Houses – upkeep		x					
Amount of open space			x				
Open space – upkeep		x					
Streets – pavement quality		x					
Streets – road quality			x				
Amount of traffic			x				
Litter – amount	x						
Mechanical noise level	x						
Public noise level	x						
Vandalism – amount	x						
Graffiti – amount	x						
Air pollution – fumes/smells	x						
Derelict	x						

⬆ **Figure 1** EQA which could be used in an urban or a rural area

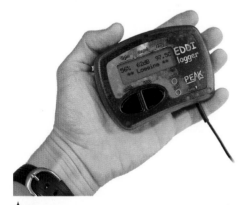

⬆ **Figure 2** A data logger. This model can record temperature, sound and light intensity

Sampling

Sample size

The reason why a sample is used, for example a sample of the population, rather than the whole is due to the amount of information that would need to be collected and handled. When deciding how to take a sample and how many samples are needed, the main consideration is how to get a true sample and how to avoid biased information that is really of no use to geography students.

A true sample has enough data to get reliable results – for example, at least five sites on a river of 10 km. For questionnaires, 50 responses is usually seen as a true sample.

Sampling techniques for some fieldwork activities

Measuring river characteristics

Where on a river would you measure its characteristics and how many times would you measure the river characteristics as it flowed from its source to its mouth?

This would depend on the size of the river but the norm would be between five and ten sites. Measurements would be equally spaced along the river's course using an OS map to work out the exact positions – although in practice this may not be possible for practical reasons, such as access to the river bank.

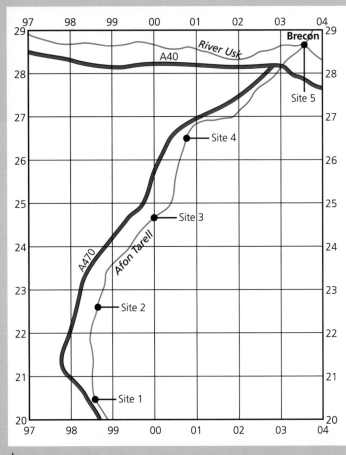

The river course is approximately 10 kilometres long, therefore five sites were chosen.

If access to the river is not possible at that particular point, another site as close to the original as possible should be used.

Study sites are equally spaced along the river's course using an OS map.

↑ Figure 3 Study sites along the Afon Tarell in the Brecon Beacons

Measuring pebble size and shape on a beach

Where on a beach would you measure the pebbles, how often and how many?

The number to measure can be worked out mathematically but it is usually accepted that ten pebbles measured every 50–100 metres, depending on the beach size and just below the high tide line, is a reliable sample site.

An example of how to measure pebble size and shape on a beach.

Measurements should be taken at regular intervals approximately 100 metres between all of the groynes.

Measurements should be taken at the same position in relation to the sea.

A quadrat should be placed on the ground; two stones should be taken from each corner and two stones from the centre at random and then their size and shape recorded.

Figure 4 Study sites along the beach between Bournemouth and Hengistbury Head, Dorset

Questionnaires

How many people should be interviewed?
The number of people to interview will depend on whether you are working alone or in a group but the minimum amount should be 50 people to get a representative sample.

Sampling methods

There are a number of different sampling methods. It is important that before a sample is completed the type of sampling method should be decided upon to give the most accurate result.

Spatial

There are three types of spatial sampling methods.

- Point sampling – this involves choosing particular points and sampling at only these points, for example specific houses down a street or places on a river.
- Line sampling – this involves taking measurements along a line sometimes called a transect, for example a line through sand dunes where vegetation is sampled every 20 metres.
- Quadrat sampling – this involves making a square on the ground and noting down what is in the square, for example the vegetation species at that particular point.

Random

This is where each member of the population has an equal chance of being chosen. It uses random number tables to decide on the sample that is taken. For example, if the random number table gave us the numbers 8, 6, 4, those would be the houses in the street that would be surveyed in a street survey.

Stratified

This is when the sample contains an equal number of results in each category. This means that the people chosen in the survey are not picked at random. For example, if a questionnaire was being carried out in a tourist resort but the views of both residents and tourists were required, the interviewer would have to ensure that they interviewed both types of people and included in the final results equal numbers of each.

Systematic

In this method, the sample would be collected according to an agreed sample. For example, every fifth person might be interviewed or every tenth house might be used in the street sample.

ACTIVITIES

1. Devise hypotheses for the following issue: 'Should a tidal barrage be built in the Severn Estuary?'
2. What would be the sequence of enquiry for a task which was investigating the presence of longshore drift on a beach?
3. Complete the following tasks for the question 'How do river characteristics change as the river flows from its source to its mouth?'
   a. Identify questions and a sequence of enquiry.
   b. Design fieldwork data collection sheets.
   c. Explain your choice of sampling method and sample size for each technique.
4. For one data collection technique you have completed, describe and explain the method.

Risk assessments

Risk assessments should be completed before students work in the field to ensure that any possible hazards have been thought about and a control put in place to minimise the risk of that hazard. Figure 5 shows an example of a risk assessment for a field trip to the Afon Tarell river in Wales. The students were studying how river characteristics change as the river moves from its source to its mouth. Information is provided about the hazards and the risks associated with those hazards.

Hazard	Misuse of equipment
Impact	Person could be injured by ranging poles if they are not used correctly.
Risk rating	6/10
Risk control	Students warned to hold poles with sharp end down in the water and to keep the poles on the bed of the river.

Hazard	Slippery rocks
Impact	Person might slip and break their ankle.
Risk rating	4/10
Risk control	Students warned about rocks before entering the river.

Hazard	Overhanging vegetation
Impact	Person could be injured in the face/eye by walking into overhanging vegetation.
Risk rating	5/10
Risk control	Students warned on bank before getting into the river about the low vegetation at this site.

Hazard	Depth and flow of river
Impact	Person could be knocked over by flow due to depth of water.
Risk rating	6/10
Risk control	Students warned on bank and taller students were told to go into the deeper areas. Students told to assess the depth with a ranging pole before moving, once in the water.

↑ Figure 5 Hazards associated with river field studies

Qualitative and quantitative data from both primary and secondary sources

It is important that you are able to use information/data from many different sources as this will be important for all parts of the examination. On all of the examination papers, you could be asked to extract and analyse information from maps, including the use of photographs and sketches in relation to maps of different scales. You could also be asked to interpret information from different sources, including fieldwork data and Geographical Information Systems (GIS). You could be asked to analyse or evaluate information of many different kinds of resources, including visual, graphical, numerical, statistical and fieldwork data. See Figure 6 for an example of an analytical comment based on evidence.

Ask yourself these questions to extract information from sources:

■ What can you see? In other words what is the overall pattern or main features?
■ Are there specific groupings of features or information?
■ Are there any anomalies or features which are particularly different?

Site	Width (m)	Depth (av)	Channel gradient (°)	Wetted perimeter
1	1.5	13 cm	7	171 cm
2	4.7	18 cm	6	540 cm
3	5.3	12 cm	5	590 cm
4	8.50	92 cm	1	10.80 cm

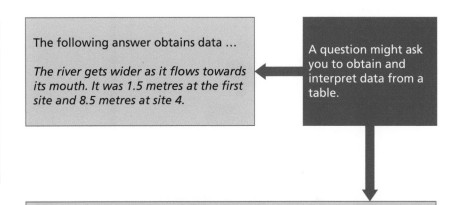

The following answer obtains data …

The river gets wider as it flows towards its mouth. It was 1.5 metres at the first site and 8.5 metres at site 4.

A question might ask you to obtain and interpret data from a table.

The following answer interprets the data …

As the river gets wider it also gets deeper; an anomaly is site 2 where it suddenly becomes deeper. This may be because of the particular point in the river where we took the measurement, which was on the outside of a meander bend.

In Paper 3, you might be given some data, a graph or other piece of evidence. The question could ask you to analyse it. Follow these points:

■ Describe what you see.
■ Ensure that you have included important facts such as dates or numbers or even names.
■ Give reasons for the patterns that you see in the data.
■ Try to link your reasons to other data that you have collected.

You could also be asked if findings you have been given support the hypothesis that has been provided.

This is a descriptive comment about the map which contains data about the counties that people came from.

The choropleth map shows where people who were interviewed came from. The people were visiting Lulworth Cove on Sunday 10th May. According to the map, 7–8 people of the ones interviewed came from Dorset and Berkshire. 1–2 people came from 10 of the counties.

Choropleth to show which county most people came from

1–2 people
3–4 people
5–6 people
7–8 people

ITALY

1. Dorset. 2. Berkshire. 3. Kent. 4. Yorkshire.
5. Lincolnshire. 6. Derbyshire. 7. Hampshire.
8. Wiltshire. 9. Essex. 10. Nottinghamshire.
11. Greater London. 12. Buckinghamshire.
13. East Sussex. 14. Oxfordshire. 15. Surrey.
16 Tyne and Wear. 17. Hertfordshire.

The choropleth map shows where people who were interviewed came from. The people were visiting Lulworth Cove on Sunday 10th May. According to the map 7–8 people of the ones interviewed came from Dorset and Berkshire. A large number of people came from Dorset because Lulworth Cove is in Dorset and therefore they did not have far to travel.

A large number of people interviewed also came from Berkshire, this is because we interviewed a lot of people who came on the trip with us.

This comment about the map contains data about the counties that people came from and analytical comments to explain the data.

Figure 6 An example of a student's work on obtaining data and interpreting a choropleth map

Electoral ward	Male (%)	Female (%)	Christian religion (%)	White ethnicity (%)	Aged 0–15 years (%)	Aged 16–64 years (%)	Aged over 65 years (%)
Abbey Meads	53	47	66	92	26	71.5	2.5
Blunsdon	51	49	77	99	17	62	21
Eastcott	53	47	61	92	15	73	12
Central	55	45	60	90	10	75	15
Shaw & Nine Elms	50	50	66	93	27	69	4
Wroughton & Chiseldon	50	50	77	98	21	61	18
St Margarets	51	49	79	97	20	63	17
Freshbrook	50	50	68	94	24	67	9
Gorse Hill & Pinehurst	51	49	64	94	22	65	13
Ridgeway villages	51	49	77	98	23	64	13

You could be asked questions that are based on the use of databases containing statistical and numerical information such as census and population data. Census data could be given and you could be asked to interpret it or to compare one ward within an urban area with another. You could also be asked to complete choropleth maps with information from the census.

Figure 7 Map of some of Swindon's electoral wards

ACTIVITIES

1 a Trace a copy of the incomplete map of Swindon's wards, Figure 7.
   b Use the data in the table above to map the percentage of white people who live in the selected wards of Swindon.
   c Use tracing paper to overlay the information on the percentage of people of working age.
2 Describe the age structure of the different wards.
3 Comment on the age structure of the wards in relation to their geographical position.

STRETCH AND CHALLENGE
· ·
Suggest reasons for the patterns shown by the data for the selected wards of Swindon.

Using digital data to obtain, illustrate, communicate, interpret, analyse and evaluate geographical information

There are many forms of digital data which you could be asked questions about in the examination. Digital data includes GIS and satellite imagery.

What is GIS?

GIS touches every part of our daily life without many people realising it. The term is daunting but the way it works isn't. It is a way of using maps digitally to make our lives easier. It allows a large amount of information to be seen by layering one set of information onto another. The base information is a map of the area being studied. It can be used on computers and on mobile devices such as tablets and mobile phones. One of the uses of GIS is global positioning systems (GPS), used by thousands of car drivers and hill walkers.

The advantages and disadvantages of GIS

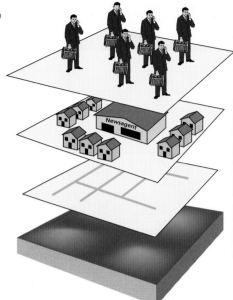

The people indicate where newspapers are delivered to each day.

This layer shows houses. The central building is a newsagent.

This layer shows the roads of the area.

Base map showing physical features.

Figure 8 A GIS map has a number of layers that can be placed on each other to build a picture of an area

Exam Tip

GIS could be used in many ways for the examination. You could be given information in the form of a GIS map and asked to extract information from it. You may be asked to analyse and evaluate information on GIS maps.

Advantages	Disadvantages
A lot of information can be seen on one map.	The information can become difficult to see if too much is put onto one map.
Information can be linked together easily to form patterns that can be analysed.	A computer or other ICT equipment is needed as the maps are digital.
GIS is available on smartphones and tablets such as iPhones and iPads.	The equipment is expensive to buy and keep up to date.
It is of great benefit to many public services such as the police and utilities.	A certain amount of training is needed to use the more sophisticated systems.
GPS has made travelling between places much easier.	

ACTIVITIES

1 What do the letters GIS stand for?
2 What is meant by the term 'layering'?
3 State four organisations or people who use GIS.
4 You order a pizza from a pizza delivery company. How could GIS be used to deliver your pizza?
5 A large supermarket chain wants to locate a new store in an area. The following layers were used to work out the best place for the store.
   a Population density
   b Land for sale
   c Transport routes
   d Competitors
   e Socio-economic groupings of residents
   Put the layers into the order that you think they would have been used. Justify your order.

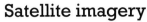

Exam Tip

When answering questions on satellite imagery, ask yourself the questions that are mentioned earlier in this chapter to extract and analyse information from written or statistical sources (see page 75).

Satellite imagery

Satellite imagery has been mentioned in Chapter 1 of this book (see page 22). However, it could be used in any part of the examination. The examination paper could provide a map with a satellite image, or part of it, and expect students to be able to interpret, analyse and evaluate geographical information from the image.

ACTIVITY

⬆ **Figure 9** A satellite image of London

Can you locate the river Thames on the satellite image?
Give reasons for your answer.

Draw conclusions from evidence about geographical questions and issues

A conclusion is when you return to your original hypotheses or questions and answer them using the evidence you have provided or that has been provided for you. The more evidence you use to back up your findings, the more plausible they will be. On Paper 3, you could be given some findings and asked to answer a hypothesis based on those findings.

Exam Tip
You should always spell accurately and write in full sentences. Don't forget to give extended arguments where required. Do not be brief and ensure that you communicate to the examiner exactly what you want to say.

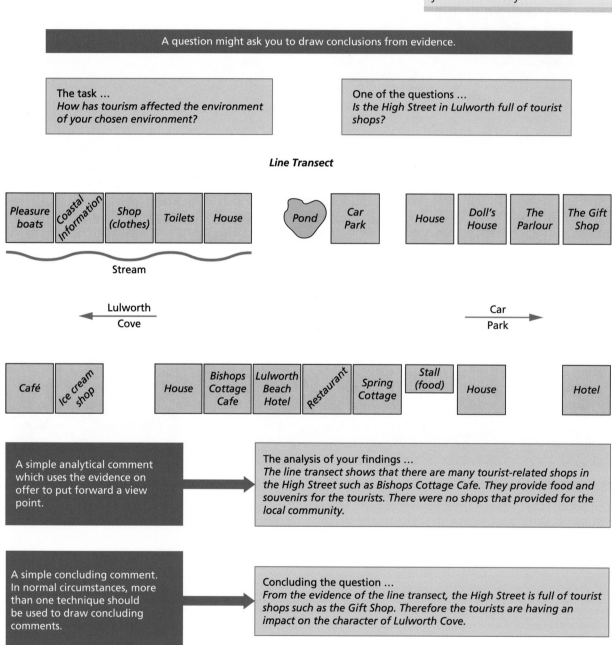

Figure 10 An example of a student's work on analysing and drawing conclusions from evidence

ACTIVITIES

1 Using your own experience in the field, state your hypotheses and write a conclusion drawing on evidence from your findings.

2 The following are the results of traffic counts taken at Lulworth Cove in Dorset. The students were trying to discover if the day of the week or time was important for the numbers and types of traffic recorded.

a Analyse the data in the tables.

b What conclusions can you draw from this evidence?

Traffic counts – Friday

Time	Out of Lulworth						Into Lulworth					
	Cars	Lorries	Coach/bus	Bike	M/bike	Total	Cars	Lorries	Coach/bus	Bike	M/bike	Total
12.30–1.00	19	9	2	0	0	30	31	8	1	1	0	41
1.00–1.30	42	0	0	0	0	42	40	2	0	0	0	42
1.30–2.00	27	0	0	0	0	27	37	2	2	0	0	41
2.00–2.30	27	1	2	0	0	30	29	2	0	0	1	32
2.30–3.00	20	1	0	1	1	23	13	0	0	0	0	13

Traffic counts – Sunday

Time	Out of Lulworth						Into Lulworth					
	Cars	Lorries	Coach/bus	Bike	M/bike	Total	Cars	Lorries	Coach/bus	Bike	M/bike	Total
12.30–1.00	107	0	3	4	3	117	65	0	5	1	0	71
1.00–1.30	108	0	3	1	1	113	75	0	1	0	0	76
1.30–2.00	131	0	0	0	2	133	45	0	0	0	0	45
2.00–2.30	131	1	0	0	4	136	56	2	2	3	0	63
2.30–3.00	82	0	0	0	1	83	69	0	1	1	1	72

REVIEW

By the end of this section you should be able to:

✔ identify questions and sequences of enquiry

✔ design fieldwork data collection sheets and collect data with an understanding of accuracy, sample size and procedures, control groups and reliability

✔ understand the need for and be able to carry out risk assessments

✔ use qualitative and quantitative data from both primary and secondary sources to obtain, illustrate, communicate, interpret, analyse and evaluate geographical information

✔ draw well-evidenced and informed conclusions about geographical questions and issues

✔ write descriptively, analytically and critically displaying good literacy skills and communicate your ideas effectively for a range of target audiences

✔ develop an extended written argument.

OS maps and questions

Chapter 5 contains section of OS maps at different scales and examination-style questions:

→ Warkworth map 1:50,000
→ Wensleydale 1:50,000
→ New Forest 1:25,000
→ Looe 1:50,000
→ Cambridge 1:50,000
→ Swange 1:50,000, 1:25,000, 1:10,000

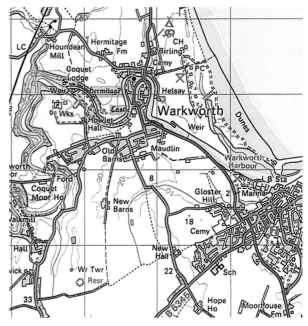

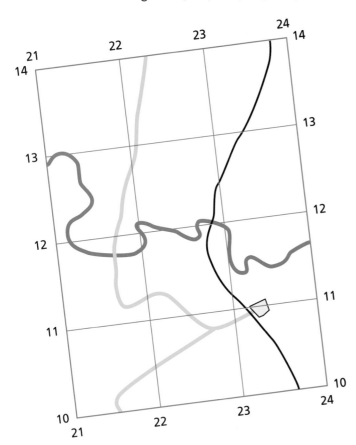

Warkworth map

1 Study Figure 1. It is a photograph taken at grid reference 248063.

   a In which direction was the camera pointing? **(1 mark)**

   b Name the river flowing under the bridge.
   i Warkworth
   ii Hermitage
   iii Aln
   iv Coquet **(1 mark)**

2 Match the coastal feature to the correct grid reference. **(3 marks)**

   Flat rock, dunes, beach.

   2507   2611   2605

3 Study Figure 2. It is a sketch map of part of the OS map extract on page 84. Trace a copy of the sketch map. Label the following features on the sketch map. You should use the correct symbol and provide a key.

   a Alnmouth station **(1 mark)**
   b Public house **(1 mark)**
   c Post Office **(1 mark)**
   d Complete the road network on the map by adding the A1068, the B1339 and the B1338. **(2 marks)**

4 Describe the course of the River Aln shown on the sketch map. **(4 marks)**

5 A family are on holiday staying at Hermitage Farm in grid square 2406. They would like to visit Warkworth for lunch and then go onto Alnwick castle in the afternoon. Describe their route. **(4 marks)**

Figure 1 A photograph of Warkworth

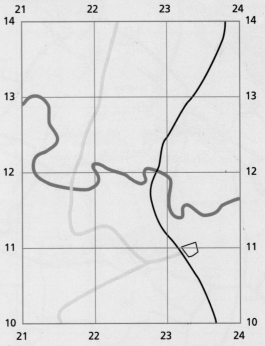

Figure 2 An incomplete sketch map of Warkworth

Warkworth 1:50 000

Wensleydale map

1 Study the OS map of Wensleydale on page 86 and Figure 3. It is a photograph taken at grid reference 876906.

   a Name the river shown in the photograph.
      i   Ure
      ii Hardraw Force
      iii Bain
      iv Brown **(1 mark)**

   b In which direction is the river flowing?
      i   east to west
      ii north to south
      iii west to east
      iv south to north **(1 mark)**

2 Copy and complete the following table by ticking (✔) in the box if Hawes (8789) and Bainbridge (9390) have that service. **(4 marks)**

Service	Hawes	Bainbridge
Public house		
Place of worship with a spire		
Visitor centre		
Post Office		

3 Identify three pieces of evidence which show that tourism is important in this area. **(3 marks)**

4 In which of the following grid squares is the gradient of the land steepest? **(1 mark)**

    9182   9385   8985

5 State the difference in height between the river valley at 881906 and the spot height at 881923. **(1 mark)**

6 Most of the settlements in the area are located in the river valleys. Suggest reasons why. Use evidence from the map in your answer. **(4 marks)**

7 Contrast the valley of the River Ure between grid reference 880905 and 900903 with the valley of Cragdale Water between grid reference 919840 and grid reference 911855. **(5 marks)**

⬆ **Figure 3** A photograph of a meander bend in Wensleydale

New Forest map

1 Study the OS map of the New forest on page 87 and Figure 4 on page 72. It is a photograph taken at grid reference 174906. In which direction was the camera pointing? **(1 mark)**

2 Give three pieces of map evidence to show that this area is important to tourists. **(3 marks)**

3 State one characteristic of the course of the River Stour in grid square 1492. **(1 mark)**

4 Describe the physical landscape of the area east of westing 16. **(3 marks)**

5 State one reason for the shape of the coastline east of westing 16. **(1 mark)**

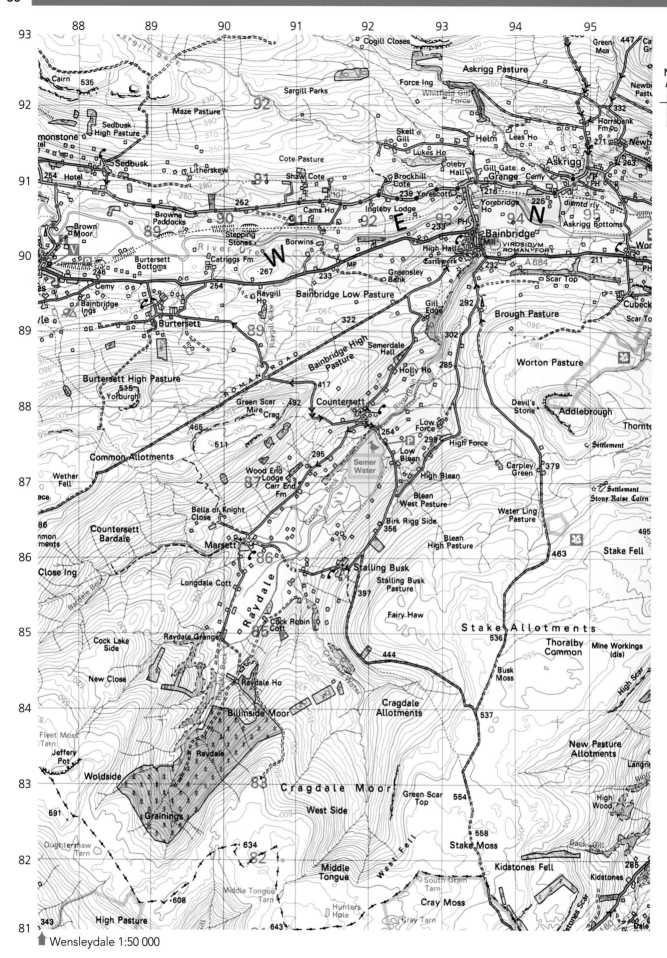

Wensleydale 1:50 000

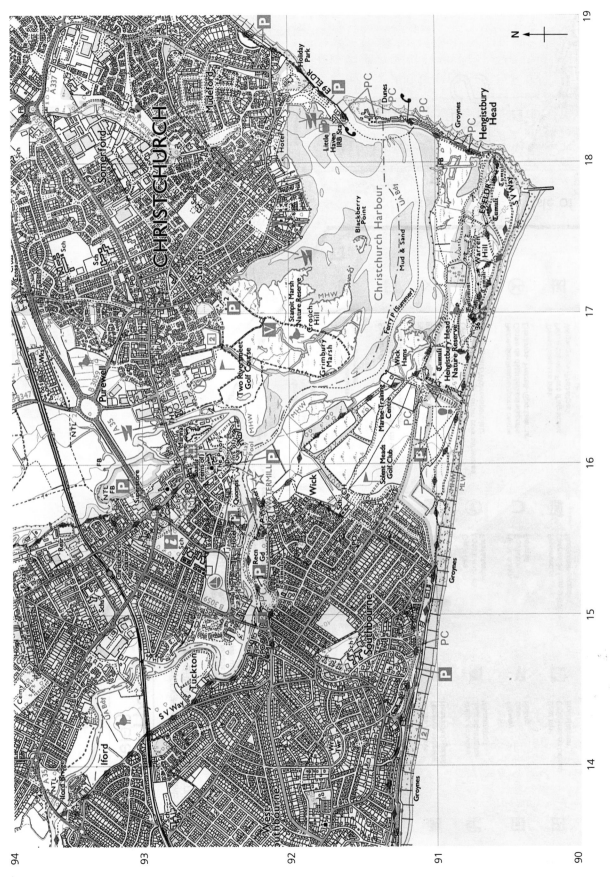

New Forest 1:25 000

Looe map

1   Study the OS map of Looe on page 90 and
    Figure 4. It is a photograph taken in grid square
    2552 facing south.
    a   What is the name of island S? **(1 mark)**
    b   What is feature T? **(1 mark)**
2   Suggest reasons for the site of Pelynt. **(4 marks)**
3   Study the OS map of the Looe area.
    a   Draw a cross-section from the triangulation
        pillar 168 in grid square 2261 to Lametton
        Mill at 260610. You may use the one in
        Figure 5 on page 89 to help you.

b   Label on the cross-section: East Looe
    River, B3245, railway line, non-coniferous
    woodland. **(5 marks)**
4   Study the OS map of Looe and Figure 6. It is a
    photograph taken in grid square 2552 facing
    north. There is information on the map that is
    not on the photograph.
    a   State three pieces of information that are
        present on the map but not the photograph.
        (3 marks)
    b   State one piece of information that is on the
        photograph and not the map. **(1 mark)**

⬆ **Figure 4** A photograph of the coastline near Looe

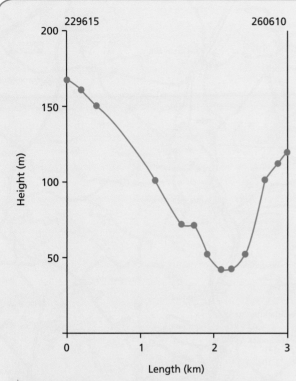

Figure 5 Cross-section from 229615 to 260610

Figure 6 A photograph of Looe

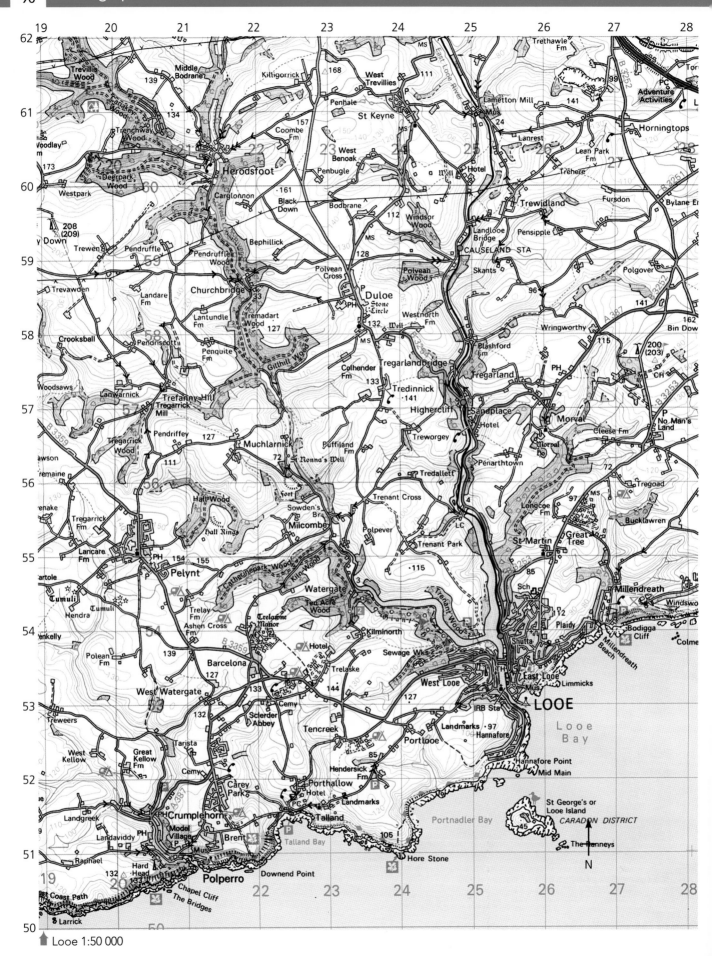

Looe 1:50 000

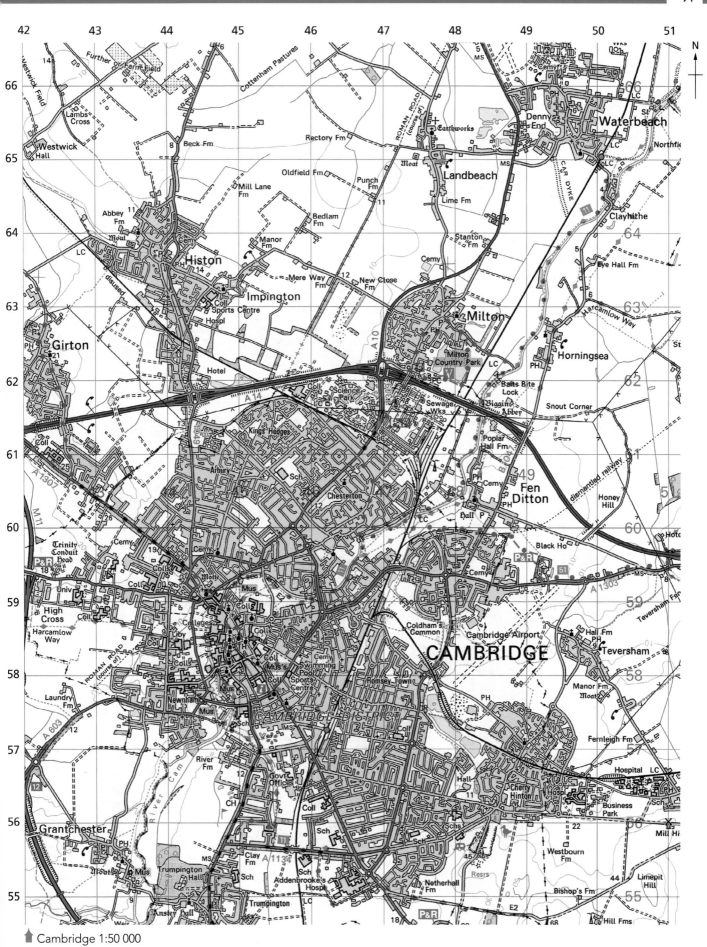

Cambridge 1:50 000

Cambridge map

1 Study Figure 7. It is an aerial photograph of an area of housing in Cambridge. Also study the OS map on page 91.

   a A number of grid squares are shown in the photograph. Choose the correct one from the following list: 4460, 4657, 4856, 4458. **(1 mark)**

   b What is the railway feature identified by the X? **(1 mark)**

   c Give the six-figure grid reference for the college identified by feature Y. **(1 mark)**

   d What is the number of the road marked Z? **(1 mark)**

   e Describe the pattern of housing identified by the red square on the photograph. Copy out the three following statements which best describe the housing identified by the red square:

   - The houses are in cul-de-sacs.
   - The housing is in a grid-iron pattern.
   - The streets have curves.
   - The houses are in blocks.
   - There are lots of deadends.
   - The streets are in a line.

   (3 marks)

2 Figure 8 is a photograph which was taken above Cherry Hinton (4856).

   a Compare the housing patterns shown in Figures 7 and 8 **(3 marks)**

   b Suggest one reason for these differences. **(1 mark)**

3 Copy and complete the table by identifying features which can be seen on the map and not on the photographs and features that can be seen on the photograph and not on the map. **(2 marks)**

	On map not on photograph	On photograph not on map
Feature 1		
Feature 2		

4 Describe the distribution of 'park and ride' sites shown on the map. **(3 marks)**

Figure 7 Aerial photograph of housing in Cambridge

Figure 8 Aerial photograph of Cherry Hinton

⬆ Swanage 1:25 000

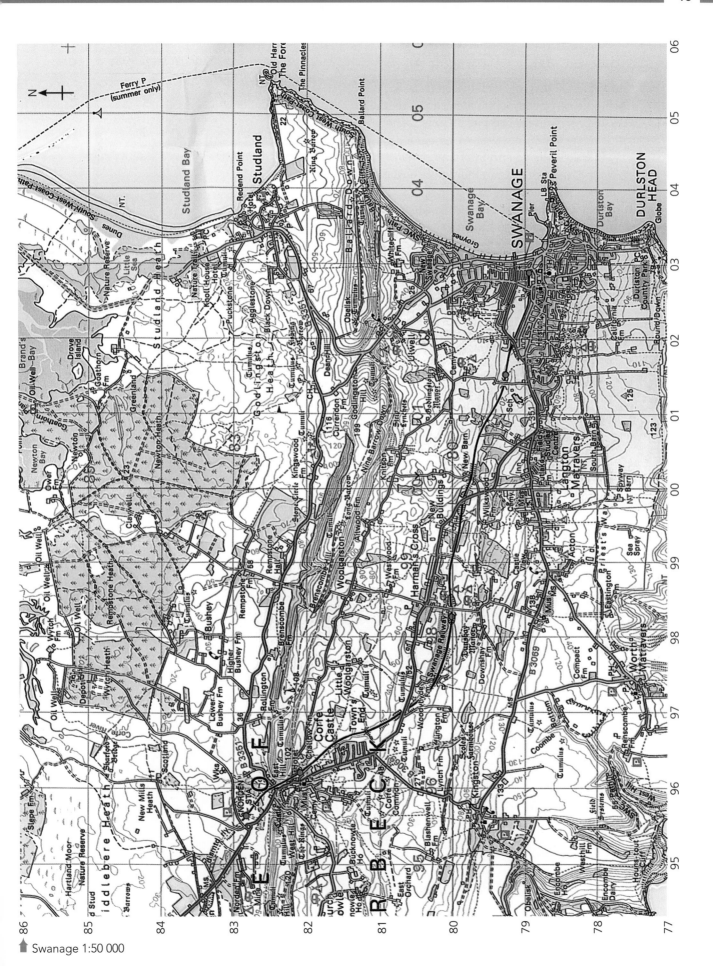

Swanage 1:50 000

Swanage map

1   Study the OS map of Swanage on page 95. A family are on holiday in the area shown on the OS map. They are staying near Corfe Castle in grid square 9582. They wish to go out for the day to visit Studland and Swanage. Answer the following questions about their route and the features they see on the way.

   a   What is the number of the road which leads from Corfe Castle to Studland?
     i   A351
     ii   B3351
     iii   B3069
     iv   A3351 **(1 mark)**

   b   They park close to the Post Office in Studland. Which one of the following is the six-figure grid reference of the Post Office?
     i   025822
     ii   038826
     iii   035822
     iv   822035 **(1 mark)**

   c   They walk in an easterly direction to Old Harry. Which one of the following owns Old Harry?
     i   Harry Thompson
     ii   National Trust
     iii   Swanage council
     iv   National Term **(1 mark)**

   d   The family leave Studland and drive to Swanage. In Ulwell they pass a tourist feature. Which tourist feature do they pass in Ulwell? **(1 mark)**

   e   They decide to return to Corfe Castle by train. Choose how many stations there are on the Swanage railway from the following:
     i   3
     ii   4
     iii   5
     iv   6 **(1 mark)**

2   Complete the sentences to compare the shape of Harman's Cross in grid square 9880 with that of Worth Matravers in grid square 9777. Use some of the terms in the box.

B3351	nucleated	junction	A351	linear

Harman's Cross is a … village. Most of the houses are either side of the main … road. The village of Worth Matravers has a more … shape with the houses being grouped around a …
(3 marks)

3   Study Figure 9. It is a photograph of Swanage Bay. Draw a sketch of the photograph. Label the following on the sketch (you will also need to refer to the OS map).
   a   the town of Swanage
   b   the beach
   c   the groynes on the beach
   d   Swanage Bay
   e   Peveril Point
   f   the pier
   (6 marks)

Figure 9 A photograph of Swanage Bay

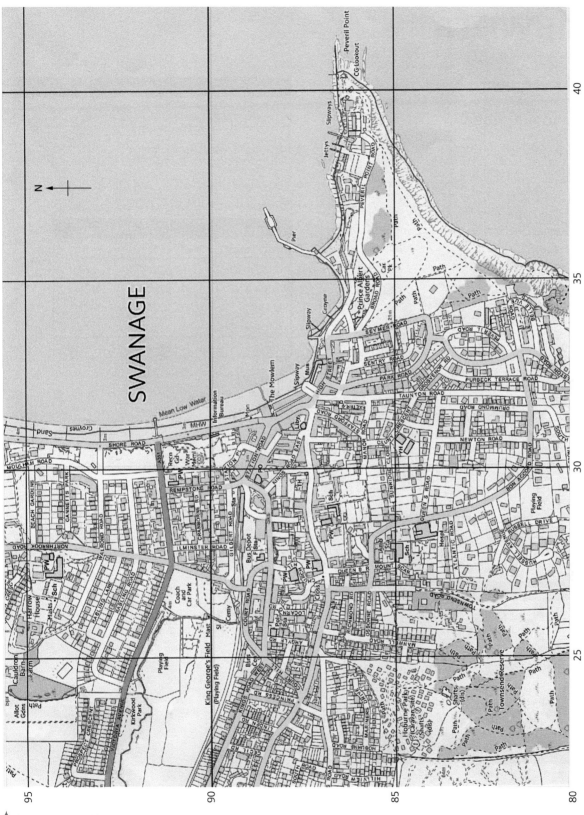

Swanage 1:10 000

Answers

Chapter 1 answers, pages 4–32

Answers to activities

Page 4

a The tropical rain forests are found either side of the Equator within the Tropics. The continent with the largest rainforest area is South America (Brazil). Europe does not have any rainforests.

b The higher land is found to the north and west of Great Britain. It is more evenly spread in Ireland. The land above 1,000 m is mainly found in Wales and Scotland with none found in the south and east of Britain. The highest land in England is in the south-west and north of the country.

c names of mountainous areas, cities

d The countries with the highest migration change are in southern Europe, including Spain and Italy. The countries with the lowest migration change are in northern Europe, including Sweden and Finland. Ireland, in north-western Europe, is an exception to this rule.

Page 5

a 50°43'N 3°31'W

b Plymouth

c A great variety of answers could be given, students working in pairs.

Page 9

1

Symbol	Feature
TH	Town Hall
☆	Viewpoint
⊞----⊞	Tunnel
⋏ ⋏ ⋏	Electricity transmission line
🌲	Non-coniferous woodland
CH	Club house

2

Symbol	Key	Four-figure reference	Six-figure reference
✝	Place of worship with tower	5683	567833
P	Post Office	5885	587857
△	Triangulation pillar	5884	582847
⌐	Golf course or links	5785	571853
☐ Home Fm	Home Farm	5583	551839
⬭	Bus or coach station	5685	568858
☐ Sch	School	5784	571841

3

Feature	Symbol
Flat rock	⎯⎯⎯

4 National Trust (always open)

5 coniferous

6 PH – Public House, P – Post Office, place of worship (church) with tower.

7

Tourist feature	Grid reference
Information	187133
Caravan site	190119
Camp site	190121
Parking	187142
(Alnwick) Castle	189138
Museum	196122

8 north-west

9 south

10 south-west

Page 11

Wensleydale

1 13.4 cm, 6.7 km

2 12.6 cm, 6.3 km

3 the river

4 0.4 km

5 10.8 cm, 5.4 km

Swanage

6 14.6 cm, 7.3 km

7 The A road distance is 10.2 cm, 5.1 km. Therefore, the B road is longer.

8 2.2 km

New Forest

9 6.5 km

10 3 km

Page 13

1 a–c See cross-section below.

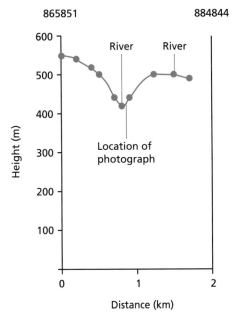

865851   884844

d south-west
e shape of the hills, small valley floor

2, 3   See cross-section below.

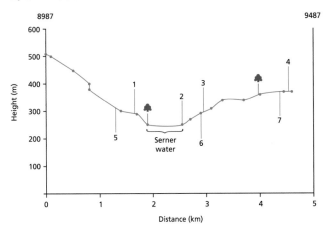

8987   9487

Page 16/17

Looe

1 The woodland is found along the river valleys such as the West Looe river. There is no woodland on the hilltops. There is no woodland above 130 m. There are some small patches of woodland, for example at 2458.

2 The railway line follows the main river valley on the map, the East Looe River. The A387 follows this river for about 3 km, then takes the route around the top of a hill to the west. Other routes mainly follow the river valleys or pass between the hills.

Swanage

3 The main settlement, Swanage, is on the coast. It is between the ranges of hills, one to the north and one to the south. There are small settlements on the slopes of the southern hills such as Worth Matravers (9777). The other larger settlements in the area are Corfe Castle, which has grown close to a gap in the northern hills that allows access to the area. There is also Harman's Cross, which is in the valley halfway between the other two settlements.

4 Straight boundaries and roadways can be seen through the woodland, which allows management. Local knowledge or map evidence provides screen for the oil field to the north.

5 There are roads on either side of the steep hill. The main A351 goes through the main valley. There are few roads to the south of the map and instead steep hills and lack of settlement.

6 a The 1:10 000 shows the most detail, less of the area is shown on the map because of the detail it contains. The 1:50 000 shows the largest area but has the least detail of that area.

b 1:50 000 ferry routes
1:25 000 leisure symbols
1:10 000 street names

c It would depend on what the trip was trying to achieve. If looking at streets in towns, the 1:10 000 would be best but if going on a coastal walk the 1:25 000 would be best. If looking at the area they are visiting, the 1:50 000 would be best.

d There are many different answers. Possible ones are below:

1:10 000	1:25 000	1:50 000
Street surveys	Going for a walk	Planning a field trip to an area
Detailed on town layouts	Doing beach surveys	Driving in an area

New Forest

7 spit
8 Expect use of key to identify features such as – sandy beach, marshland, shingle, lowland.
9 1491 is very man-made – groynes on beach, built-up area, some cliffs. 1690 is very open and rural, nature trail, low lying.
10 meander bend

11 It is a meander bend, river approximately 75 m wide, built up within 0.5 km of river, the rest is left as floodplain although built-up area closer in some places, golf course on eastern bank.

Page 18

1 Post Office, public house, place of worship with a spire, milestone, road

2

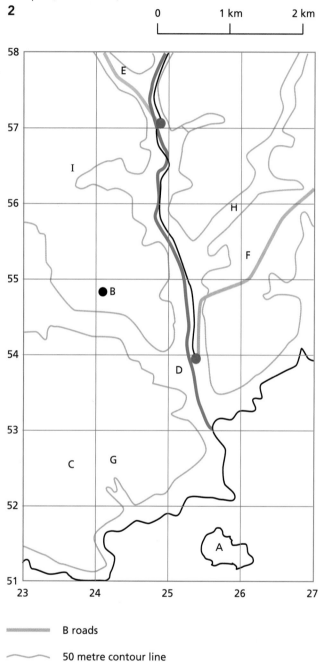

3

Place	Feature on map
Island A	St George's or Looe
Spot height B	115 m
Tourist attraction C	Caravan and camping site
Tourist attraction D	Parking
Road number E	B3254
Road number F	B3253
Settlement G	Port Looe
Type of woodland H	Non-coniferous
Farm I	Puffiland Farm

Page 21

1

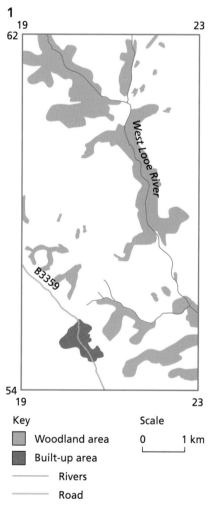

2 Answer should be a simple sketch with the following marked:
  a the built-up areas of Alnmouth and Warkworth
  b the Rivers Aln and Coquet
  c the coastline and all coastal features
  d any tourist information features in the area

Page 25

1

Ground photo	Oblique aerial photograph	Vertical aerial photograph	Satellite photograph
Shows a small area in detail.	Shows a larger area but less defined.	Shows a smaller area but more detail.	Shows a large area of the coastline.
Can clearly see beach and type of vegetation.	Shows outline of coastline, hills and lowland areas.	Does not show if the area is flat or hilly. Coastline shown clearly.	Can distinguish beaches and landforms to some extent.
Shows groynes on the beach and houses in the distance.	Shows position of settlements, wooded areas and fields.	Shows areas that are fields and areas that are built up. Shows actual houses, roads/street patterns.	Can identify built-up areas, field patterns and wooded areas.

2 Human features: castle, defensive site, village in the centre of the meander bend, large fields around the village. Physical features: flat land, no sign of animals so probably arable (crop-farming), wide bend (meander) of the river, coastline with sandy beach in the background, which is possibly used by tourists in the summer.

Page 27

  a north-west
  b

Letter on photograph	Grid reference	Feature
V	Six figure = 254535	Number of road = A387
W	Four figure = 2454	Name of woodland = Trenant wood
X	Six figure = 257530	Identify symbol = Beacon
Y	Four figure = 2554	Name of river = East Looe River
Z	Six figure = 251537	Identify symbol = Parking

  c Many answers are possible: cars parked, the actual buildings, wooded areas around housing to west of river that are not on the map.
  d Many answers are possible: museum, road numbers, names of villages.

Page 30

1

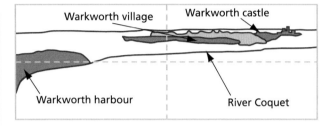

2

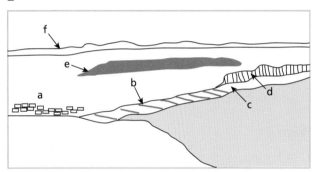

Page 32

1 The population pyramid should be set out like the one on page 32 with men on one side and women on the other. There should be five blocks on each side.

2 Example annotations could be: the dip in the age range 15–24 could be because it is a smaller age range than the others; the larger numbers on the female side after the age of 55 are because, on average, women live longer than men; there are more males up to the age of 55.

3 The age ranges denote children, middle-aged and retired people, with the working population split into young adults and main working age. The age ranges are easier to work with than bands of 10 years although they do hide the detail.

4 17.2 per cent

Chapter 2 answers, pages 38–57

Answers to activities

Page 38

1   population pyramid accurately drawn
2   There are large numbers of people in the 20–29 age group: 15 per cent males and 12 per cent females. There are very few people over the age of 69: only 13.5 per cent.
3   It clearly shows whether any of the age groups has more of the population than the others. The data is continuous.
4   flow lines
5   Pedestrian counts are flows of people, therefore flow lines are the appropriate technique.
6   All of the countries have a winter maximum. The amount that they have in summer is very low, the highest being 80 mm in Rome. The lowest summer total is in Tripoli with 1 mm. The highest winter total is Tel Aviv, the lowest is 125 mm in Tripoli. Tel Aviv has the greatest difference between summer and winter precipitation, Rome the smallest.

Page 44

1   Advantages: shows at a glance the peaks and flows in the traffic, comparisons between times are easy to make, easy to construct, continuous data. Disadvantages: can be a problem if the number of vehicles varies greatly between the times, would be better if displayed as a flow because it doesn't show direction, which a flow would.
2   a   flow line map
    b   It shows direction as well as numbers of vehicles.
    c   Many more people went into Lulworth on Sunday than any other flow. The peak flow was 136 vehicles between 2 p.m. and 2.30 p.m. The lowest flow is out of Lulworth on Friday p.m. between 3 p.m. and 3.30 p.m., when ten vehicles left Lulworth.
3   Arrows should go along the streets where the people walked. There is no scale on the key. The arrows are different lengths. It is not clear from the location of the arrows where the counts were actually completed. The number of pedestrians should not be written on the arrows, it should be determined from the key.

4   map redrawn taking the points in question 3 into consideration
5   histogram
6   compound line graph drawn as in the 'how to' box on page 41

Page 45

1   pie diagram as in Figure 11
2   The number employed in primary industry has declined from about 8 per cent in 1991 to about 2 per cent in 2006. The number in secondary industry declined between 1991 and 2001. It was then fairly stable between 2001 and 2006. Tertiary industry has increased over the whole period from about 63 per cent in 1991 to 85 per cent in 2006.
3   compound bar chart

Page 47

1   The pictogram could be pictures of people but better if pictures are of different vehicles.
2   bar chart
3   Pictures of vehicles show clearly what has been counted. Numbers of vehicles can be seen easily by the key.
4   Data is more accurately presented. The chart is much easier to construct.

Page 49

1   choropleth drawn
2   Most tourists go to France with 81.9 million tourists visiting there. The tenth most popular world tourist destination is Mexico with 21.4 million tourists. This is the least to expect. There should be other comments about countries within these figures.
3   Advantages: shows immediately at a glance the pattern made by the data, visual representation is easy to complete.
    Disadvantages: only shows a spread of data not an amount, the ranges of data used can make the map ineffective in showing the pattern.
4   bars on a map of the world
5   The bars show immediately by their height the amount; with a choropleth the key would need to be referred to. The choropleth does not show a specific amount but a range of data.
6   A histogram is better for continuous data.

Page 51

1   map completed
2   The greatest number of tourists went to Spain with 14 million visitors. The least number of tourists went to Portugal, Greece, the Netherlands and Turkey with only 2 million. The only destination in the top ten that is not in Europe is the USA with 4 million tourist visits.
3   Advantages: shows clearly the differences due to the different sizes of the circles; shows location so the spread of the data can be seen at a glance. Disadvantages: difficult to construct; data is shown in a range not an actual value.

Page 53

1   scatter graph drawn
2   There is a positive correlation.
3   New Zealand, Algeria
4   If a country has piped water to houses, more water will be used. It is a percentage therefore that will alternate with the other uses, which are agriculture and industry. If the country is a dry one, the agricultural percentage will be higher.

Page 55

1   The majority of the cities are located in Asia. There are three cities in South America and three cities in North/Central America. The only city in Europe is Paris. There are no cities with over 10 million inhabitants in Australia.
2   Advantages: show distribution, very visual, can see at a glance where cities are clustered. Disadvantages: can get clustered, is time-consuming to construct.
3   A table of continents could be constructed with the names of the cities listed under the appropriate continent.

Page 57

1   dispersion graph completed
2   Pebbles are larger at site 1. The largest pebble at site 2 is still as large as the middle range of pebbles at site 1. The smallest pebble is 3 mm at site 2 but site 1 has a pebble only 2 mm bigger. Therefore, there is little difference between the sizes.
3   The pattern shown is that the smaller pebbles are found at the second site. The larger pebbles are found at the source of the river.
4   scatter graph
5   It would also show the dispersion of the pebbles.
6   The dots on the graph could have been a certain shape. Therefore, they could have drawn a pictogram of the shape of the pebble on the graph where the size was located.

Chapter 3 answers, pages 60–66

Answers to activities

Page 60

$1500 \times 1200 = 1,800,000$ square metres (m²)

Page 63

1 a 290
  b 314

2 a

Site	Mean	Mode	Median
1	7.4	7.75	7.75
2	7.5	7.5	7.5
3	6.2	6	6
4	6.5	6.5	6.5
5	6.45	6	6.25
6	4.5	4.5	4.5
7	3.4	3	3.25
8	3.5	2	3.25

b

Technique	Advantages	Disadvantages
Median	• Easy to calculate • Not distorted by extreme results	• Does not take into consideration precise values
Mode	• Easy to calculate	• Not very useful if the sample size is small
Mean	• Uses all the data so is very inclusive	• Extremes can be a problem if the sample size is small

Page 64

1 Width of 6.6 m
   Width of 8.3 m
2 250 per cent

Page 66

On average, the further from the groyne, the smaller the **pebble size.**

Chapter 4 answers, pages 73–81

Answers to activities

Page 73

1 Hypotheses such as:
 The barrage would have a detrimental effect on fish stocks in the area.
 The barrage would have a negative visual impact on the area.
2 Devise techniques to test for the presence of longshore drift, carry out the field work, collate results, present the data, analyse, conclude and evaluate your work.
3 a Hypotheses/questions such as:
  The width of the river will increase as it flows from source to mouth.
  The velocity of the river will increase as it flows from source to mouth.
  See sequence of enquiry on page 69.
 b These will be individual to the student but should include location of survey(s), room to write down a number of techniques at each site, and space for repetition of sampling.
 c Again, individual responses but must include explanation.
4 Answers will vary depending upon technique chosen.

Page 77

1 Map completed
2 The highest percentage of working age is in Swindon Central. The highest percentage over 65 is in Blunsdon. The highest percentage under 16 is in Abbey Meads.
3 The 'younger' population is in the centre of Swindon. The older population is on the edge of Swindon.

Page 78

1 Geographical Information System
2 Layering is the ability that GIS gives you to put different types of information on a map by overlaying on a computer.
3 ambulance service, supermarket chains, public utility companies, police
4 A satellite navigation system is GIS; this would be used by the delivery person to find your address.
5 land for sale, population density, socio-economic groupings of residents, transport routes, competitors
 Comments such as: The land has to be first because otherwise they cannot build the store. The others could be in a different order but if no one lives there then there is no point in putting a shop there.

Page 79

Answers such as: The dark colour across the centre of the photo in a bendy line indicates water. The Isle of Dogs meander bend can be seen clearly on the satellite image.

Page 81

1 student's own answer
2 a Analyse means to describe and explain. For example, Sunday was a busier day than Friday with a total of 327 vehicles going into Lulworth whereas on Friday only 169 vehicles went into Lulworth, a difference of 158 vehicles. This may be because Sunday is the weekend and fewer people are at work or school.
 b Conclusions may be that Lulworth is much busier at the weekend than in the week because many more vehicles came and left the car park.

Chapter 5 answers, pages 83–96

Answers to exam-style questions

Page 83

Warkworth map

1 a south
   b iv River Coquet   (2 marks)
2 2611 flat rock
   2507 beach
   2605 dunes   (3 marks)
3 See sketch map for answers.   (5 marks)

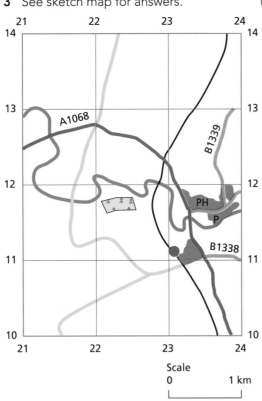

Scale
0   1 km

4 The river valley is at a height of 10 m. The river is in
   its lower course near the sea. The river has many
   (meander) bends. A railway line crosses the river in
   grid square 2212. In grid square 2311, the river is
   crossed by the A1068. The river flows to the south
   of Lesbury. There is a small coniferous woodland to
   the south of the river in grid square 2211.   (4 marks)
5 Leave Hermitage Farm, turn left onto the
   secondary road. After approximately 1 km, arrive at
   a junction with the A1068. Turn left onto the A1068
   towards Alnmouth, drive for approximately 5 km.
   Go straight across the roundabout in grid square

2311, continuing on the A1068. After approximately
5 km, you will then reach the town of Alnwick.
Continue through the town on the B6341, parking
in the car park close to the castle.   (4 marks)

Page 85

Wensleydale map

1 a i River Ure   (1 mark)
   b iii west to east   (1 mark)
2

Service	Hawes	Bainbridge
Public house	✓	✓
Place of worship with a spire	✓	
Visitor centre	✓	
Post Office	✓	

                                                        (4 marks)

3 Any three tourist features from the map key: visitor
   centre, craft centre, and so on.   (3 marks)
4 9182   (1 mark)
5 230 m 535 m = 305 m   (1 mark)
6 The settlements are in the valleys because
   that is where the main road is (the A684). The
   settlements are not on the hills because they
   are too steep. This is shown by the contour lines
   being close together. The settlements are found
   in the river valleys to have easy access to water for
   drinking.   (4 marks)
7 The River Ure bends. It has a wide flat valley floor
   about 0.5 km. The river flows from west to east.
   There is a caravan site to the south of the river. It
   is approximately 220 m above sea level. Cragdale
   water has a number of smaller bends than the
   Ure, which has one big one in the section we are
   looking at. It has steep sides and very little flat
   land in the valley bottom. The river flows from
   south-east to north-west.   (5 marks)

New Forest map

1 west   (1 mark)
2 beach 1391, parking 1691, nature trail 1690/1691,
   nature reserve 1790, PC 1691   (3 marks)
3 meander bend, floodplain, levee   (1 mark)
4 river estuary, headland, spit, mud and sand,
   marshland   (3 marks)

5 There is a headland with a spit which has been formed due to Longshore drift occurring along the coastline. (1 mark)

Page 88–89

Looe map

1 S = St. George's or Looe Island
T = Flat rock (2 marks)
2 It is a wet point site. It has a spring in the village, which would have supplied water to the villagers. It is above the floodplain of the river. It is on the junction of a number of secondary roads. (4 marks)
3 one mark for each label and one mark for drawing the cross-section

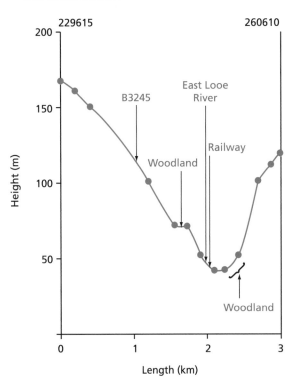

(5 marks)

4 a museum, IRB station, information centre, place of worship with a tower (3 marks)
b boats, benches, promenade (1 mark)

Page 92 Cambridge map

1 a 4657 (1 mark)
b railway station (1 mark)
c 472572 (1 mark)
d A1134 (1 mark)
e The housing is in a grid-iron pattern. This means that the housing is in blocks. The streets are in a line and meet each other at right angles. (3 marks)

2 a cul-de-sacs, curves, not in a straight line compared to comments in 1e (comparative comment not required, implicit comparison is satisfactory) (3 marks)
b The housing is on cheaper land at the edge of the city. (1 mark)

3

	On map not on photograph	On photograph not on map
Feature 1	Churches	Fields
Feature 2	Road numbers	Types of buildings

(2 marks)

4 The 'park and ride' sites are all on the outskirts of Cambridge. They are close to major roads that lead into the city. There is a 'park and ride' site in grid square 4259. This is close to a motorway junction. (3 marks)

Page 96

Swanage map

1 a B3351 (1 mark)
b 035822 (1 mark)
c National Trust (1 mark)
d Camp and caravan site (1 mark)
e 5 (1 mark)
2 Harman's Cross is a linear village. Most of the houses are either side of the main A351 road. The village of Worth Matravers has a more nucleated shape with the houses being around a junction. (3 marks)
3 Sketch (1 mark), accurately shading the town of Swanage (1 mark), beach and groynes (1 mark) and Swanage Bay (1 mark), Peveril Point (1 mark), the pier (1 mark). (6 marks)

Index